Green Energy and Technology

Climate change, environmental impact and the limited natural resources urge scientific research and novel technical solutions. The monograph series Green Energy and Technology serves as a publishing platform for scientific and technological approaches to "green"—i.e. environmentally friendly and sustainable—technologies. While a focus lies on energy and power supply, it also covers "green" solutions in industrial engineering and engineering design. Green Energy and Technology addresses researchers, advanced students, technical consultants as well as decision makers in industries and politics. Hence, the level of presentation spans from instructional to highly technical.

Indexed in Scopus.

Indexed in Ei Compendex.

Xiaolin Wang
Editor

Future Energy

Challenge, Opportunity, and, Sustainability

Editor
Xiaolin Wang
School of Engineering
University of Tasmania
Hobart, TAS, Australia

ISSN 1865-3529 ISSN 1865-3537 (electronic)
Green Energy and Technology
ISBN 978-3-031-33905-9 ISBN 978-3-031-33906-6 (eBook)
https://doi.org/10.1007/978-3-031-33906-6

This Springer imprint is published by the registered company Springer Nature Switzerland AG
The registered company address is: Gewerbestrasse 11, 6330 Cham, Switzerland

Paper in this product is recyclable.

Conference Committees

Advisory Committees

- Vladimiro Miranda (IEEE Fellow), INESC TEC and University of Porto, Portugal
- Nikos Hatziargyriou (IEEE Fellow), National Technical University of Athens, Greece
- Tony Wood, Grattan Institute, Australia

Conference Chair

- Xiaolin Wang, University of Tasmania, Australia

Program Chairs

- Michael Negnevitsky, University of Tasmania, Australia
- Hee-Je Kim, Pusan National University, South Korea
- He Cai, South China University of Technology, China

Steering Committee Chair

- Lin Chen, Chinese Academy of Sciences, China; Tohoku University, Japan

Local Organizing Committees

- Ibrahim Sultan, Federation University, Australia
- Farhad Shahnia, Murdoch University, Australia

Technical Committees

- Vignesh Vicki Wanatasanappan, Universiti Tenaga Nasional, Malaysia
- Vladimir Simón Montoya Torres, Universidad Continental, Perú
- Winda Nur Cahyo, Universitas Islam Indonesia, Indonesia
- Fadoua Tamtam, National School of Applied Sciences, Morocco
- Nhlanhla Mbuli, University of Johannesburg, South Africa

Preface

The 2023 7th International Conference on Sustainable Energy Engineering (ICSEE 2023) is dedicated to issues related to Sustainable Energy Engineering. It was supposed to be held in Sydney, Australia, from February 17 to 19, 2022, but changed to be held online as most authors were unable to attend the on-site meeting due to the pandemic and travel restrictions.

The Technical Program Committee put together an excellent program, which includes presentations from two Keynote Speakers – Prof. Tony Wood (Fellow to the Australian Academy of Technology and Engineering, Grattan Institute, Australia) and Prof. Nikos Hatziargyriou (FIEEE, National Technical University of Athens, Greece), and four Plenary Speakers – Prof. Michael Negnevitsky (University of Tasmania, Australia), Prof. Mohammad Rasul (Central Queensland University, Australia), Prof. Firoz Alam (RMIT University, Australia) and Assoc. Prof. Wenming Yang (National University of Singapore, Singapore). The program includes more than 20 papers that are accepted for presentations and publication in the ICSEE 2023 conference proceedings.

ICSEE 2023 offered a great opportunity for discussions of methods, technologies, systems, and practices in different areas of sustainable energy engineering among researchers from various places around the world. The conference enabled participants to exchange ideas with other researchers and build connections upon which international collaborations can stand firm and thrive, and aim at solving civil engineering and architecture problems, at present and in the future.

The proceedings of ICSEE 2023 cover a range of topics, including, but not limited to, Grid-Connected Renewable Energy Systems and Renewable Energy Applications, Building Energy Efficiency, Energy Management and Thermal Comfort, Power System and Electric Power Load Management, Energy-Saving Technology and Thermal Management, Fuel Consumption, Transportation Carbon Emissions and Climate Change Management, etc.

We would like to thank all the authors that submit their papers to the conference. We would also like to thank sincerely all our colleagues who extended their help in terms of organizing, reviewing, promoting, etc. for their continuous help and support, and we promise you all that we will continue our efforts to make the joint ICSEE to be one of the best conferences.

Efforts taken by peer reviewers contributed to improving the quality of papers and provided constructive critical comments, improvements and corrections to the authors are gratefully appreciated. We are very grateful to the international/national advisory committee, session chairs and administrative assistants who selflessly contributed to the success of this conference. Also, we are thankful to all the authors who submitted papers, because of which the conference became a story of success. It was the quality of their presentation and their passion to communicate with the other participants which made this conference a grant success.

As you read these proceedings, please plan to submit a proposal to the ICSEE 2024 conference, which is scheduled for February 2–4, 2024, in Brisbane, Australia. We are looking forward to seeing you at ICSEE 2024!

Warmest regards,

Hobart, TAS, Australia Xiaolin Wang

Contents

About the Editor

Xiaolin Wang, Ph.D., is a Professor in the School of Engineering, University of Tasmania. He worked as a Research Fellow in the Department of Mechanical Engineering at the National University of Singapore from 2002 to 2005. In 2003, he was a Visiting Scientist at the University of Siegen, Germany, funded by DAAD, Germany. In 2005, Dr. Wang joined the School of Mechanical and Chemical Engineering at the University of Western Australia. In 2012, he joined the School of Engineering, University of Tasmania, where he is now a Professor and Deputy Director of the Centre for Renewable Energy and Power Systems. His research interests include cooling and power engineering, energy storage and conversion, desalination, and utilization of renewable energy. Dr. Wang is a Fellow of Engineers Australia and was named a leader in the field of Thermal Science by *The Australian's Research Magazine* in 2018. His national and international reputation in thermal science is evidenced by his achievements and awards, including the DAAD visiting fellowship by DAAD Germany in 2003 and the Ludwig Mond Prize 2005, by the Institute of Mechanical Engineers (IMechE), the Australian China Young Scientist Exchange Program award in 2009, and the Australian Japan Emerging Research Leader Program award in 2016 by the Australia Academy of Technological Sciences and Engineering (ATSE). He received the Dean's Award for outstanding research performance in 2016. He is a subject editor of *Applied Thermal Engineering*, associate editor of the *Proceedings of the Institution of Mechanical Engineers, Part E: Journal of Process Mechanical Engineering*; *Scientific Reports;* and *Frontiers in Built Environment*, topic editor of *Applied Sciences; Energies; Sustainability;* and *Thermo*, and an editorial board member for five other international journals. He has completed many national/international research projects with a total value of more than $7 million. He is also a member of the Blue Economic CRC ($329 million). He has published more than 240 international journal and conference papers.

Part I
Grid-Connected Renewable Energy System and Renewable Energy Application

Chapter 1
A Design and Fabrication of an Automated Solar Energy Tracker Integrated Four-Sided Reflector-Based Box-Type Solar Cooker to Increase Efficiency by Absorbing Maximum Solar Energy

Ashik Mahmud, Md. Sayeduzzaman, Shahrukh Islam, Touhidul Hasan, Rafid Hasan, and Mohamed EL-Shimy

1.1 Introduction

New energy sources are needed to address the global energy crisis, especially in emerging nations. Solar energy shows potential. Globally, fossil fuels provide most of the energy. Environmental issues boost the importance of renewable energy, and a survey of local fuel consumption confirms it. Rural areas use firewood (75%) and cow dung cake (25%) for cooking purposes [1]. Burning wood deforests and decertifies, and firewood can create health issues. Renewable cooking energy sources are gaining popularity in underdeveloped nations. Solar energy is a free and clean energy solution. Solar cookers replace wood and manure. Solar cooking conserves energy and cooks meals [2]. A solar box cooker is like a hotbox. We can cook our food without any cooking gas, kerosene, electricity, coal, or wood. Solar energy is a clean, reliable, and abundant form of energy that can be harvested

---

A. Mahmud (✉)
Rajshahi University of Engineering & Technology, Rajshahi, Bangladesh

M. Sayeduzzaman · T. Hasan
American International University-Bangladesh, Dhaka, Bangladesh

S. Islam
BRAC University, Dhaka, Bangladesh

R. Hasan
Bangladesh University of Engineering & Technology, Dhaka, Bangladesh

M. EL-Shimy
Ain Shams University, Faculty of Engineering, Electrical Power and Machines Department, Cairo, Egypt
e-mail: mohamed_bekhet@eng.asu.edu.eg

X. Wang (ed.), *Future Energy*, Green Energy and Technology,
https://doi.org/10.1007/978-3-031-33906-6_1

to provide electricity and heat [3]. S. Mahaver et al. (2012) present the design development and thermal and cooking performance studies of a novel solar cooker; it is named as Single Family Solar Cooker (SFSC) [4]. A tiltable box-type solar cooker was designed and fabricated to meet the cooking needs of a typical family of five members [5]. A four-sided reflector-based design would allow for a more efficient and effective way to absorb maximum solar energy by using four-sided reflector panels to direct the sunlight onto the cooking area. Some authors previously tested a dual-axis solar cooker. Solar casseroles were made and recipes were tested in this casserole-type cooker [6]. The hotbox solar cooker works best in summer. Sunlight heats solar cookers. Easy, safe, eco-friendly solar cooking. A solar oven with four reflectors and an automated tracker had tested.

The remaining seven portions were in order. Section 1.2 gives the overall literature review of the research, which outlines the approach. Section 1.3 describes the design and modeling of the prototype. Section 1.4 shows how to implement and use the prototype. Section 1.5 shows the study and data analysis. Section 1.6 outlines potential research advancements. Section 1.7 lists the benefits and applications of this research. Section 1.8 shows the study's conclusion.

1.2 Literature Review

Modern cooking uses much energy. Cooked food is mostly eaten. According to FAO, this industry generates 30% of the world's energy. Inefficient cooking with biomass pollutes indoor air [7]. Prof. Viral K Pandya et al. (2011) began their study to analyze the performance of box-type solar cookers under Gujarat climate conditions in mid-summer to improve cooker workability. A review of box-type solar cooker applications and designs is given here. Two cookers were evaluated. The first kind has a black base, and the second has coal. These designs were tested in fixed and tracking modes. The fixed cooker's thermal efficiencies ranged from 25.2% to 53.8% about 12–1 pm, with an average overall efficiency of 32.3%. The thermal efficiency of a sun-tracking black coal cooker ranged from 28% to 62%, with an average of 43.8% [8]. Mahaver et al. (2016) discuss a solar cooker's design, development, heating, and cooking performance (SFSC). Sayeduzzaman Md. et al. published a research article where they used automated solar tracker to charge batteries from solar energy to run minimal loads during the power outage [9]. This cooker's strengths are small, practical construction, hybrid insulation, and polymeric glazing. Solar cookers use one-sided reflectors. Four-sided reflectors with automated sun trackers are used to maximize cooking output.

1.3 Design and Modeling

This portion deals with modeling, designing, and materials selections for the proposed solar cooker. With the help of theoretical efficiency, the area and dimensions of the reflectors and the cooker are determined. The governing equation of a solar cooker is

$$\begin{aligned} &\text{Efficiency} = \text{Output energy/Input energy} \\ &\eta = mC\text{p}\left(T_{\text{w}_2} - T_{\text{w}_1}\right)/\Delta t * I * A \end{aligned} \tag{1.1}$$

where

m = mass of water
Cp = specific heat of water
Δt = duration of water heating
I = solar radiation in terms of W/sq. m
A = inside surface area of the cooker

From the above equation for producing 100 °C temperature and taking the efficiency of the cooker as 33%, the inside surface area of the cooker will be

$$\begin{aligned} A &= mC\text{p}\left(T_{\text{w2}} - T_{\text{w1}}\right)/\Delta t^* I^* \eta \\ &= 3^*4200^*\ (100 - 25)\ /3600^*700^*.33 \\ &= 1.14\ \text{ sq.m} \\ &= 1764\ \text{ sq.in} \end{aligned}$$

1.3.1 Specified Dimension of the Box

The computations determined the area required to construct the box by considering the box's length and width, which is 15″ × 12.5″ = 187.5 sq. in. designed as shown in Fig. 1.1, where the length of the box is 15″ and the width of the box is 12.5″.

Fig. 1.1 Prototype of the box design

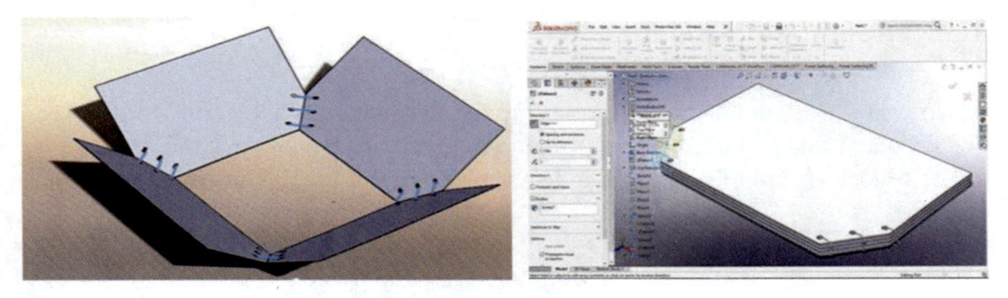

Fig. 1.2 Prototype of reflector design

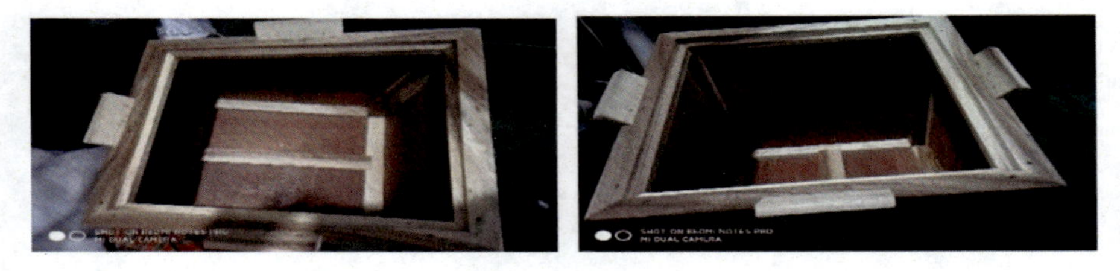

Fig. 1.3 Box of solar cooker

1.3.2 Specified Dimension of the Reflector

Calculations were made, and corner-cutting parts were determined using the reflector's length and width values. The length of the reflector is 24″, the width of the reflector is 18″, the area of the reflector is $4 \times (24'' \times 18'')$ sq. in., and the area of eliminated portion is $4 \times (4 \times 8)$ sq. in. So, the final area of the reflector is $4\{(24'' \times 18'') - (4'' \times 8'')\} = 1600$ sq. in.

A prototype of the reflector has been designed in AutoCAD, shown in Fig. 1.2.

1.3.3 Fabrication of the Solar Cooker

The wooden box was assembled. The wood was cut with a hacksaw and finished by a finisher. The upper part includes three reflector sheets supplied, shown in Fig. 1.3.

The board was trimmed to size, sanded, and glued to trap the foil. The plywood reflector was coated in aluminum foil shown in Fig. 1.4.

The inner portion of the box was filled with cotton around 1.5″ with glue (first figure) and then coated with aluminum-wrapped sponge wood (second figure). Cut sponge wood was coated in aluminum foil with adhesive shown in Fig. 1.5.

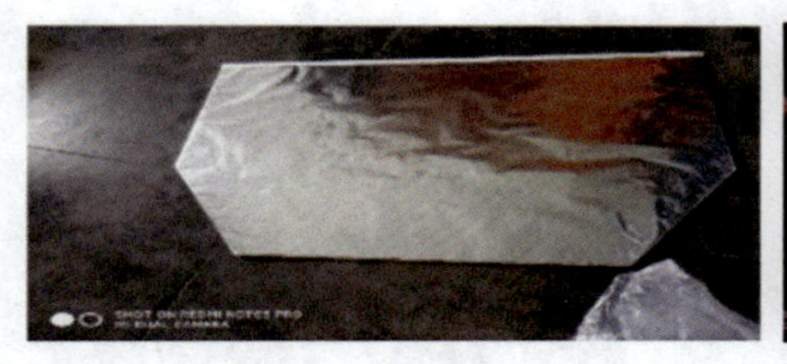

Fig. 1.4 Reflector of solar cooker

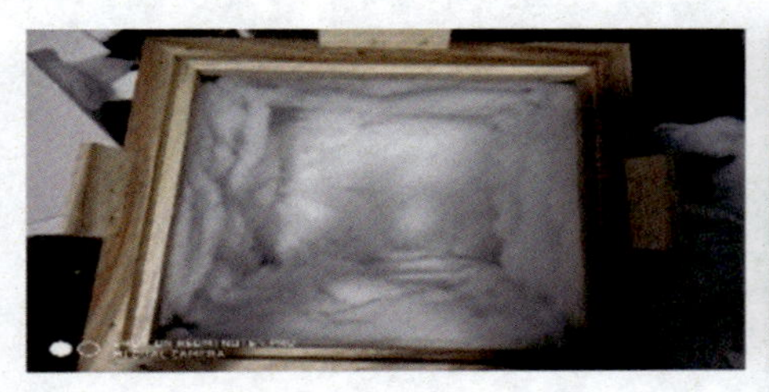

Fig. 1.5 Box with insulating material

1.3.4 Materials Used for Fabricating Four-Sided Box-Type Solar Cooker

- Plywood: Plywood does not react dramatically to extreme temperatures like other materials.
- Glass piece: Glass is an inorganic material that is fused at high temperatures and cooled rapidly, which is transparent and allows light and heat to pass through it. Glass has the property of trapping heat.
- Aluminum foil: Aluminum foils thicker than 25 μm are impermeable to oxygen and water. The reflectivity of bright aluminum foil is 88%, while the dull embossed foil is about 80%.
- Black absorber pot: A black absorber pot absorbs sunlight very effectively.
- Insulating material (cotton): Foam-type cotton has high insulating properties.
- Thermometer: The temperature scales in °C and ranges from 0 to 120 °C.
- Arduino Uno and LDR: For programming and running the automated solar tracker system using light-dependent resistors (LDRs) [10–11].
- Servo motor: Futaba S3003 is a standard servo motor that is mainly used.

1.4 Implementation of the Prototype and Working Procedure

1.4.1 Experimental Setup

Solid works was used to build the prototype per the specifications. Figure 1.6 shows a glass serving plate was placed on top of the black ceramic pot within the box.

Fig. 1.6 Design setup

Fig. 1.7 Implementation of the designed prototype

Another glass plate topped the box. Our designed reflector sheets were placed on the box. Last setup: correctly positioned reflector sheets, glass plate, and absorber pot. Following the prototype setup, the final configuration was completed. Correctly positioned reflector sheets, glass plate, and absorber pot. The solar tracker system was integrated into the cooker, which was positioned to face the sun for maximum sunlight as shown in Fig. 1.7. The projected prototype and automated procedure require 12 V power.

1.4.2 Working Procedure

Sunlight Concentration A surface with a high specular reflection conversed sunlight on a small cooking area. Depending on the surface's geometry, the efficiency may vary.

Conversion of Light Energy to Heat Energy Solar cooker receives light from the sun. The receiver pan then absorbs the heat. This conversion is maximized by using a solar tracker and materials that absorb and hold maximum heat, like any black-colored pot.

Trapping Heat Energy Isolating the air inside the cooker reduces convection heat transfer. The greenhouse effect that a glass lid creates on a pot improves heat retention and reduces convection loss. Glass lids also increase light absorption [12].

Fig. 1.8 Schematic diagram based on the principle of box-type solar cooker

Gain the Most Intensity The schematic diagram of the developed and implemented prototype is shown in Fig. 1.8. It uses an LDR programmed in an Arduino Uno using the Arduino IDE to produce more heat, which is probably indeed more significant than 100 °C between 11:00 and 12:00.

Software-Based Implementation In the software part, the code is constructed in C programming and inserted in Arduino IDE for automated solar tracking system.

1.5 Results and Data Analysis

From 11:00 am to 12:00 pm, a cooker's performance was tested every 30 min. 2:00 pm is summer's optimal temperature. The cooker's performance, including merit and efficiency, was calculated using initial and final temperature, mass, chamber dimension, and collector area. Seven-day experiment results varied. This section describes the experimental results. Solar intensity, gloomy weather, and other factors reduced the experimental value. As with a solar cooker, the more intense the heat on the absorber, the higher the efficiency. Sunlight influences solar radiation. Midday was when tests showed the most efficiency. The reflector and cooker get the same amount of solar energy on experiment day; instantaneous efficiency and figure of merit were calculated. Data experiment. Solar radiation is affected by the sun's position. The designed cooker reached to 113 °C and 30.26% efficiency. The existing collector was 64 °C and 23.43% efficient with a tracking system. Figure 1.9 shows the upward-sloping temperature difference vs. the daytime curve, which shows daytime temperatures climb.

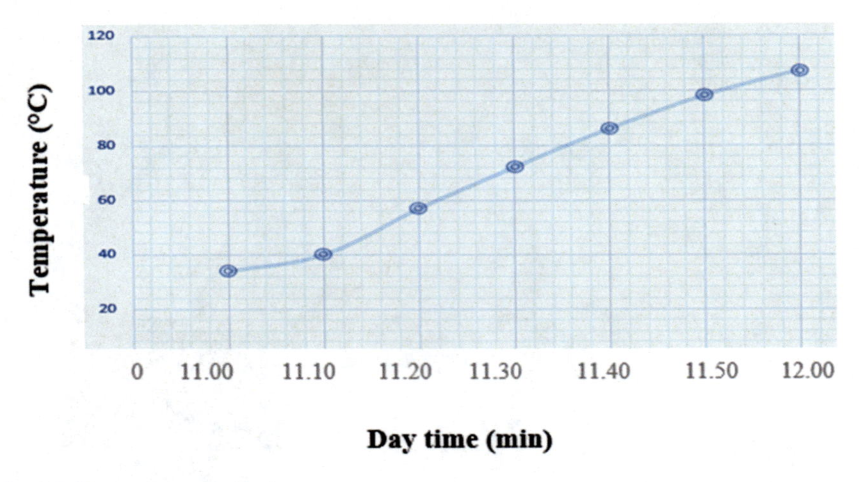

Fig. 1.9 Temperature vs. daytime curve

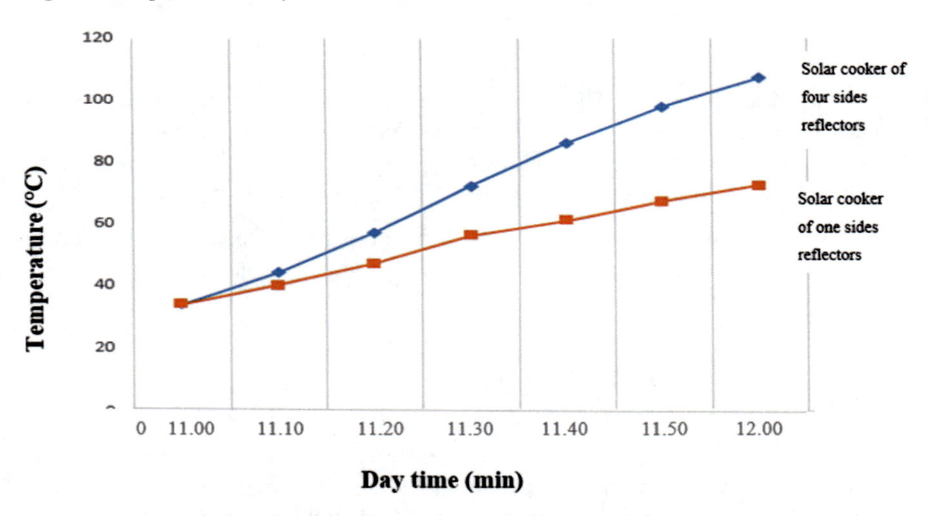

Fig. 1.10 Comparison of temperature difference vs. time curve

Figure 1.10 Temperature rises over time. The cooking chamber temperature was raised to 113 °C each day in 1 h. Using a four-sided reflector with an automated solar tracking system to maximize sunlight caused a linear temperature rise. The 7-day average 1-h intensity was 750 W/m^2.

The temperature rises of a newly designed automated solar tracker integrated four-sided reflector-based box-type solar cooker, and an earlier one-sided reflector-based solar cooker is compared in Fig. 1.10 and Table 1.1. The highest temperature for sun cookers designed for use was 113 °C, whereas, for one-sided reflectors, it was 64 °C.

Table 1.1 Comparison of day vs. maximum temperature difference

Day	Maximum temperature recorded for one-sided cooker	The maximum temperature recorded for four-sided cooker	Solar intensity (W/m^2)
01	62	111	747
02	61	110	741
03	64	113	750
04	62	111	747
05	59	106	735
06	60	109	740
07	64	113	750

1.6 Future Developments

The solar cooker was constructed following the plan. Instead of using aluminum foil as a reflector, an aluminum reflector sheet can produce far superior results. Instead of regular glass, the low-tampered iron glass may transmit and trap the most heat. The solar cooker's performance will be improved by completely enclosing the cooking chamber in a highly insulating material. Future research could add a light-dependent resistor for nighttime security. LDRs can turn lights on and off in the dark.

1.7 Advantages and Applications

It has various benefits. i.e. It's easy to install and produce, minimal power source needed; almost no running and maintenance costs. It is reasonably priced with electrical energy. Simple equipment makes it easy to get the desired temperature. Save money and fuel. Environmentally friendly, secure, and with no ash. Reduced CO_2 emissions to reheat any food. For boiling rice, eggs, and water. Make a cake, pudding, etc., to make alcohol. The prototype is employed for cooking items like coffee, tea, and other things.

1.8 Conclusion

With the help of an automated solar tracking system, a solar cooker in the form of a wooden box was successfully built. The cooker harnessed the sun's absolute power throughout the day. The cooker's adaptable design makes it possible to fold, unfold, and move it from one place to another. The components of a four-faced reflector were used, and by positioning them at the right angle, they were incorporated into the tracking system. The maximum efficiency in this project was 30.26%, which was approximately 7% higher than the existing solar cookers available on the market.

Additionally, the maximum temperature was obtained at 113 °C, approximately 50 °C higher than the existing one.

Acknowledgment First and foremost, the authors wanted to thank the Almighty and their parents. For their encouragement and support, the authors would like to thank Asma-Ul-Husna, Assistant Professor, Department of Mechanical Engineering, Rajshahi University of Engineering & Technology (RUET), Vice Chancellor Dr. Carmen Z. Lamagna, and the Electrical and Electronics Engineering Department of American International University-Bangladesh. Finally, the authors would like to thank BRAC And BUET's Vice Chancellors for their inspiration and motivation for the research.

References

1. Sahin, A.D., Dincer, I., Rosen, M.A.: Thermodynamic analysis of solar photovoltaic cell systems. Sol. Energy Mater. Sol. Cells. **91**, 153–159 (2007)
2. Negi, B.S., Purohit, I.: Experimental investigation of a box type solar cooker employing a non-tracking concentrator. Energy Convers. Manag. **46**(4), 577e604 (2005). https://doi.org/10.1016/j.enconman.2004.04.005
3. Soni, P., Chourasia, B.K.: A review on the development of box type solar cooker. Int. J. Eng. Sci. Res. Technol. **3**, 3017–3024 (2014)
4. Mahavar, S., et al.: Design development and performance studies of a novel Single Family Solar Cooker. Renew. Energy. **47**(2012), 67–76 (2012)
5. Al-Nehari, H.A., et al.: Experimental and numerical analysis of tiltable box-type solar cooker with tracking mechanism. Renew. Energy. **180**(2021), 954–965 (2012)
6. Cuce, E., Bali, T.: Improving performance parameters of silicon solar cells using air cooling. In: Fifth international edge energy symposium and exhibition, Denizli, Turkey; 27–30 June 2010
7. Pandya, V.K., et al.: Assessment of thermal performance of box type solar cookers under Gujarat climate condition in mid summer. Int. J. Eng. Res. Appl. **1**(4), 1313–1316 (2011)
8. Algifri, A.H., Al-Towaie, H.A.: Efficient orientation impact of the reflector of box type solar cooker on the cooker performance. Solar Energy. **2**(70), 165–170 (2002). https://doi.org/10.1016/s0038-092x(00)00136-5
9. Sayeduzzaman, M., et al.: Design and implementation of an automated solar tracking system to run utility systems at minimal loads during load-shedding by charging solar batteries Advances in Transdisciplinary Engineering. IOS Press (2022). https://doi.org/10.3233/atde221224
10. Arduino Uno. Components101, components101.com/microcontrollers/arduino-uno. Accessed 9 Dec 2022
11. mybotic, and More by the author: "LDR Sensor Module Interface with Arduino." Instructables, www.instructables.com/LDR-Sensor-Module-Users-Manual-V10. Accessed 10 Dec 2022
12. "Solar Cooker – Wikipedia." Solar Cooker – Wikipedia, 4 June 2014, en.wikipedia.org/wiki/Solar_cooker

Chapter 2
Design and Implementation of an Automated Hybrid Sustainable Energy Generation from Earth-Battery and Solar PV System

Md. Samiul Islam Borno and Md. Abdur Rahman

2.1 Introduction

The soil is a capacitive element that enables the potential charge to be stored under each restricted soil cell in presence of natural water. In an Earth-Battery, the potential energy of the soil in a restricted space is transformed into electrical energy. Under the restricted cells, the vast amount of electrons remains stable under the soil components so, for generating DC power from the soil, the creation of the artificial electromagnetic field (AEMF) is responsible for the initial movements of electrons and use of the components as easily obtainable elements such as soil, cell container, connecting wires, anode, cathode, and natural water were necessary. It is known that the potential difference is 0 on the open ground, but the soils were restricted under individual many cell containers from the open ground so that the electrodes internal charge particle properties could force the electrons to move under the restricted soil cell area. The anodes and cathodes are employed as dipole electrodes to create this initial force that enables to move the electrons under each soil cell. In a soil cell container, carbon**6** is applied as an anode, and aluminum**13** as a cathode. Since the anode pulls electrons toward it and the cathode pushes them away, the attract-repel charge state of the electrodes could produce initial movements of electrons into a soil cell; as a result, there generates a stable amount of usable DC power. In a single soil cell, the electrodes placed that maintained a maximum distance between each other so that one soil cell can produce maximum voltage from a cell by following Coulomb's law. Since water is a conductive substance, adding water to this soil cells makes easier for electrons to move around under each cell and lets us improve how well each soil cell conducts electricity. Each soil cell can produce 1 V (DC) [1].

M. S. I. Borno (✉) · M. Abdur Rahman
Department of Electrical and Electronics Engineering, Faculty of Engineering, American International University-Bangladesh, Dhaka, Bangladesh
e-mail: samiul.borno@gmail.com

X. Wang (ed.), *Future Energy*, Green Energy and Technology,
https://doi.org/10.1007/978-3-031-33906-6_2

Consequently, Earth-Battery can produce a significant amount of DC power from this project that can be used for the hybrid renewable energy model, and the hybrid project might serve as a rooftop renewable energy generation unit to provide enough power to the household appliances.

Everyone is aware that a power crisis is imminent and concerning that everyone must work together to find the best way to use the world's economical and renewable energy resources. The main concern of this project is to make an alternative solution for now and in the future by slowly fixing the problems that are causing the electricity crisis. This method generates green energy for ecological development and also saves money because the soil is easily accessible and can be used as a renewable energy generation unit for green building infrastructure that may help to improve a healthier city and more sustainable environment. This paper highlights how such a hybrid renewable energy system can be constructed.

The remaining ten sections had been arranged in order. Section 2.2 provides the overall literature review of the study, and Sect. 2.3 represents the methodology and the modeling. Section 2.4 discusses the working procedures of the designed system. Section 2.5 presents a software-based simulated implementation, and Sect. 2.6 shows the hardware implementation. Section 2.7 explains the experimental results and data analysis, and Sect. 2.8 offers the possible visual applications of this research. Section 2.9 discusses future developments. Finally, Sect. 2.10 describes the conclusion of the research study.

2.2 Literature Review

The optimal way to obtain green energy from natural resources could be a smart hybrid solar power system combined with a sustainable Earth-Battery. There are several types of hybrid system research being investigated. M.S.I. Borno et al. researched a project called Sustainable Earth-Battery that can produce sufficient energy that can be used in the hybrid renewable energy systems [1]. Jing et al. researched battery lifetime enhancement via smart hybrid energy storage plug-in modules in standalone photovoltaic power systems [2]. Badwawi et al. reviewed a paper on systems that use both solar PV and wind energy [3]. Wadi et al. published a case study on a smart hybrid wind-solar street lighting system [4]. Meskani et al. proposed an intelligent hybrid energy source based on solar energy and batteries [5]. Deshmukh et al. proposed a model for hybrid renewable energy systems [6]. Nema et al. published a review on a hybrid energy system using wind and PV solar [7]. Khalil et al. researched on hybrid smart grid with sustainable energy-efficient resources for smart cities [8]. Sayeduzzaman, M., Borno, M.S.I., et al. published a research on a method and design for developing an IoT-based auto-sanitization system powered by Sustainable Earth-Battery [9]. This article describes the process, design, and strategy for developing such a system. It could lead to a better way to get renewable energy, and it could also lead to creative solutions.

2.3 Methodology and Modeling

This hybrid energy system combining solar power with Earth-Battery is a system for producing green energy from renewable resources. Although the method for producing power from solar energy is well understood, the mechanism for the Earth-Battery is more exciting. It is noteworthy that the Earth-Battery can produce as much voltage as necessary by increasing the number of soil cells, but there is a limitation to how many amperes can be produced since the soil cell configurations matter. Figure 2.1 illustrates the MATLAB-Simulink model of cell arrangements to get a usable amount of power, and Fig. 2.2 shows the hardware design of the soil cells arranges in two stages.

The soil cells in this system are made to raise the minimum amperage needed to combine the two different types of renewable DC energy sources. There are a total of 48 soil cells in both series and parallel configurations, and each soil cell can produce 1 V (DC) with 20 mA, and two backup cells can be used in the project

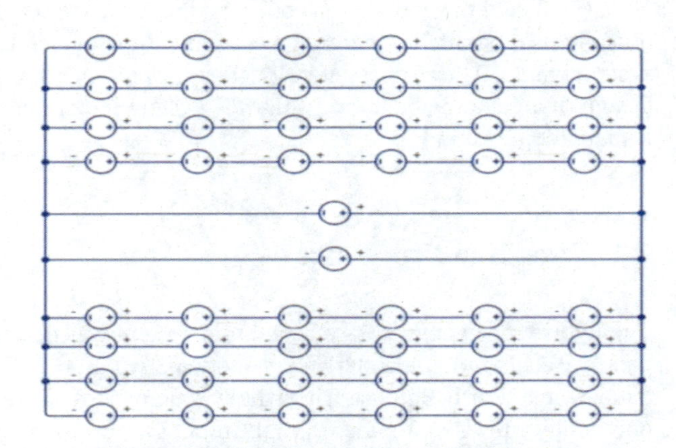

Fig. 2.1 Simulink model of cell arrangement

Fig. 2.2 Hardware of soil cell arrangement

Fig. 2.3 3D model design of a renewable energy generation unit on the rooftop

design to maintain a constant voltage level. The soil cells organized as series-parallel configurations so that the output voltage and current remain at constant 12 V (DC) rating, where tried to keep the same as the solar battery's output voltage rating of 12 V (DC), and the connections between the series-connected cells are organized as four times in parallel to increase a constant amperage and to keep the whole hybrid system stable. The primary goal of this research is represented by a 3D animated design of an automated hybrid solar PV system integrated with a sustainable Earth-Battery, depicted in Fig. 2.3.

2.4 Working Procedure

The hybrid renewable energy generation system was designed with a 20-Watt solar panel, PV charge controller, relay module, Arduino UNO, LDR sensor, primary battery, and Earth-Battery. The whole system is designed under two conditions; one is day time and another is night time. During the day, the sun's lights hit the solar panel and turned them into electricity and the electricity is then stored in a primary battery. Earth-Battery can simultaneously power low-power devices such as DC lights and mobile charging. As Earth-Battery is a water-activated battery, it also has the capability to get naturally recharged by itself [1]. When night falls, the microcontroller receives a signal from the LDR sensor and then microcontroller switches the relay module to allow the primary battery, and Earth-Battery can combine their power and supply power to the output loads. A block diagram of the designed prototype is shown in Fig. 2.4.

There are two scenarios in which this hybrid renewable energy system can be utilized. Figures 2.5 and 2.6 illustrate the system operations during the day and at night.

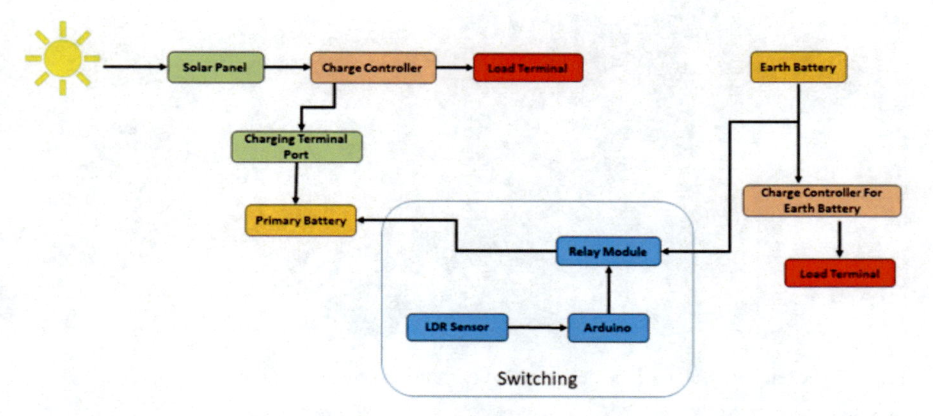

Fig. 2.4 A block diagram of the hybrid renewable energy generation system prototype

Fig. 2.5 Daytime operation of the system

Fig. 2.6 Nighttime operation of the system

2.5 Software-Based Simulated Implementation

A block diagram had been developed for the project simulation. The system is designed to improve system's efficiency as a product of extensive research. The

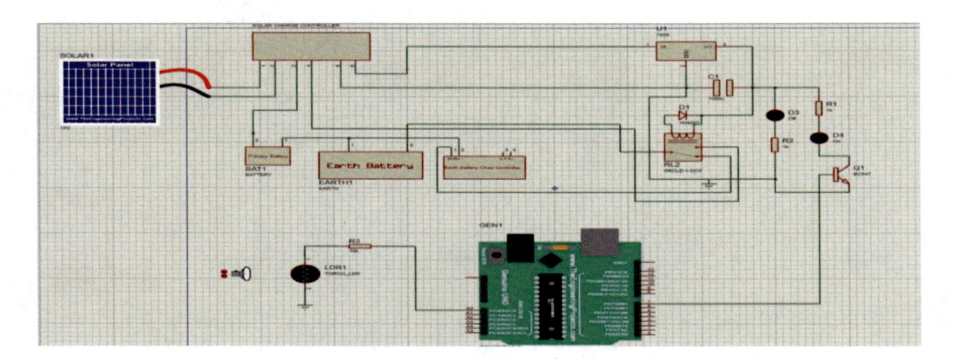

Fig. 2.7 Schematic design of the designed hybrid renewable energy model prototype

solar battery could only store power for a certain amount of time and is usually used to power home appliances. For this reason, the goal of this hybrid energy model is to supply the output loads to last longer. Two different operating conditions were maintained to maximize the system's efficiency. The LDR sends a signal to the microcontroller to turn on the relay, and then the two renewable sources might be combined and can provide power to the output DC loads by indicating day and night. Using Proteus, a software-based simulation has been developed based on the schematic diagram presented in Fig. 2.7.

2.6 Hardware Implementation

In the hardware design, a 20-Watt solar panel had used with a PV charge controller. For building the Earth-Battery, needs to process the soil mixed with natural water and two dipole electrodes that were implanted into each soil cell. Then the Earth-Battery requires the arrangements of soil cells in both series and parallel configurations to maintain a stable DC voltage and current rating for combining the two DC renewable sources. That could produce a combined DC output power that shown in Figs. 2.8 and 2.9, which demonstrates the hardware project modeling of the hybrid renewable energy system.

2.7 Experimental Results and Observation Data Analysis

This section presents the experiment that was practically observed in this research study as well as examined the experimental results of the designed hybrid prototype.

Figure 2.10 displays the process on how the hardware prototype operates during the day (OFF-load condition). The primary lead-acid battery (12V - 38Amp) is charged during this phase using solar energy, which takes around 6 h to get fully

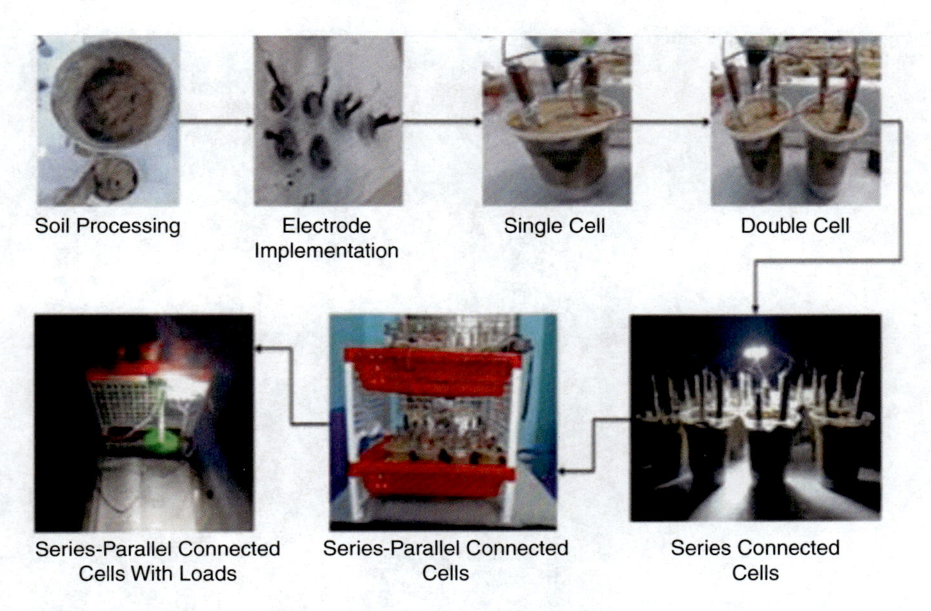

Fig. 2.8 Earth-Battery making process [1, 9]

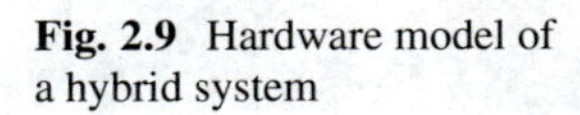

Fig. 2.9 Hardware model of a hybrid system

charged. Figure 2.11 demonstrates the hardware prototype for nighttime (ON-load condition) operation. The hybrid system can provide power to the same 120 W loads for over approximately 5 h during this phase, while the primary battery requires around 4 h to completely deplete.

The solar primary battery's initial fully charged practical voltage is displayed in Fig. 2.12 at 11.27 V (DC), whereas the Earth-Battery's initial fully charged voltage is shown in Fig. 2.13 at 11.26 V (DC). Because the voltages of the two sources are almost identical, a hybrid system can be designed by combining the two DC sources.

From Fig. 2.14, it is shown that only Earth-Battery's charging and discharging capabilities with load applications respect to the time of observations and observed

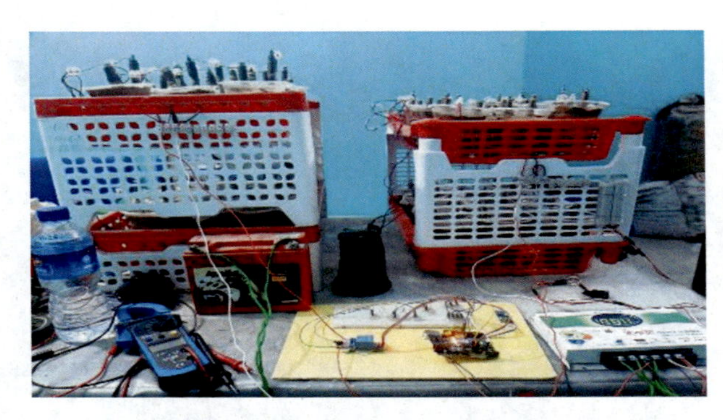

Fig. 2.10 Daytime operation of hardware

Fig. 2.11 Nighttime operation of hardware

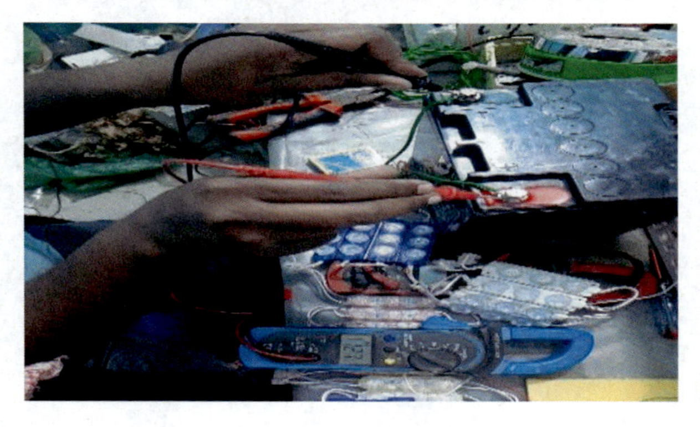

Fig. 2.12 Primary battery voltage observation

that the Earth-Battery has the capability to get naturally recharged. The Earth-Battery could be naturally recharged to its maximum capacity in 2–3 h without any external power source—this project's most outstanding feature [1]. There is no need for any input power to charge the Earth-Battery, and also Earth-Battery can supply a usable amount of DC power to the applied loads by combining with this hybrid renewable energy system.

Following the foregoing, Fig. 2.15 shows a graph that indicates a primary battery's timing analysis with respect to V (DC). Based on this analysis of the data, a primary battery takes almost 6 h to fully charge and 120 W DC loads take 4 h

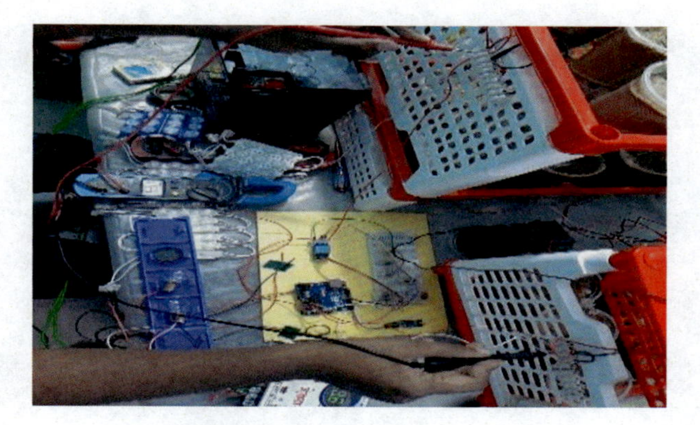

Fig. 2.13 Earth-Battery voltage observation

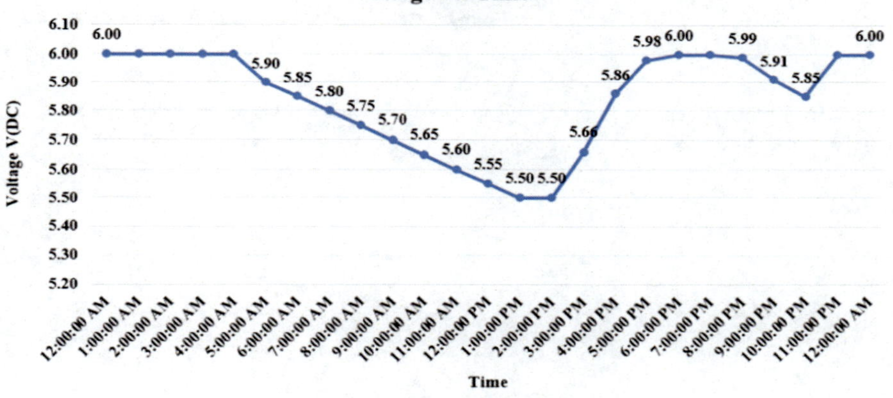

Fig. 2.14 Earth-Battery's **voltage vs. time** observations for charging and discharging [1, 9]

to fully drain the battery. A graph showing the charging profile timing analysis of the hybrid renewable energy system concerning V (DC) is shown in Fig. 2.16. The important finding of this research is that the hybrid system could provide DC power for approximately an hour longer than the primary battery, according to the practically observed data analysis.

2.8 Visual Possible Applications

This research approach may have some novel applications that can benefit the human society. In order to ensure that human civilizations can live in healthy environments, Fig. 2.17 demonstrates a visual representation of various possible uses of the hybrid renewable energy model that can be applied to both urban and rural locations.

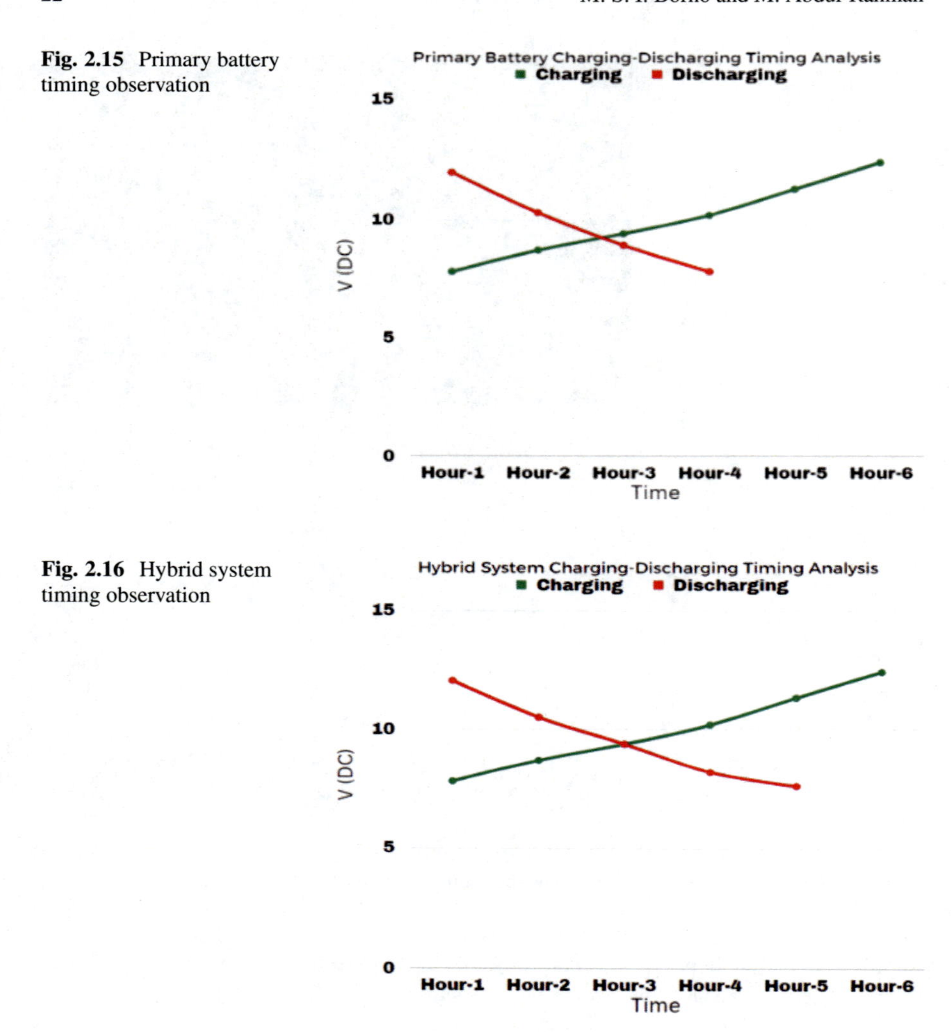

Fig. 2.15 Primary battery timing observation

Fig. 2.16 Hybrid system timing observation

2.9 Future Developments

Hybrid energy system functionality can always be updated. Earth-Battery efficiency can be increased by reconstructing the cell design procedure with more research. Earth-Battery charge controller can be designed to increase the ampere. The number of solar panels and soil cells can be readily increased to get the required power that depends on applications.

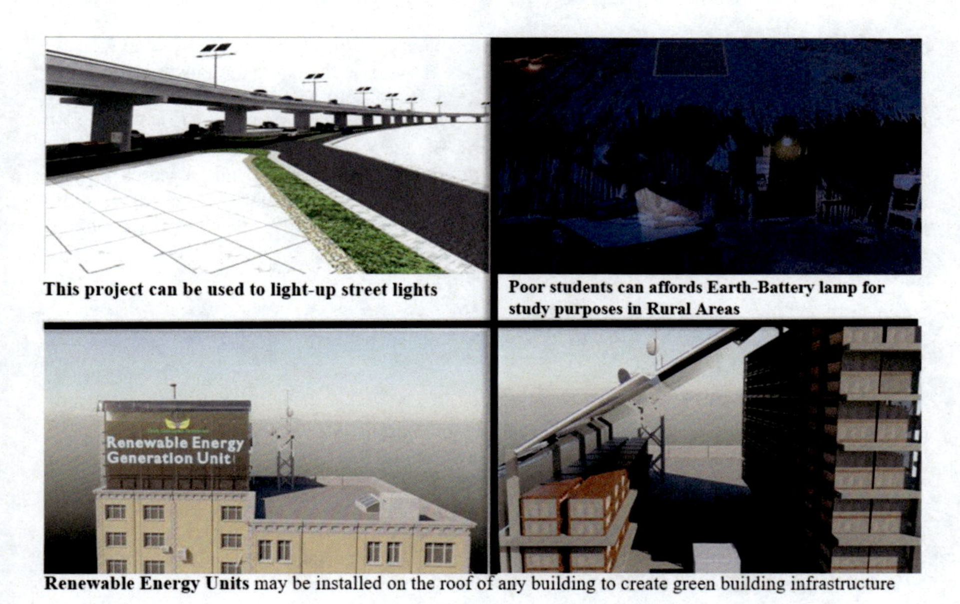

Fig. 2.17 Possible visual graphics for various innovative applications of the hybrid energy model

2.10 Conclusion

This research recommends a feasible method for obtaining green energy from natural resources through a functioning hybrid prototype. This proposed project would deliver combined renewable energy in an effort to extend the output load supply time to the conventional solar power system. It is also conceivable to extend the hybrid system's output load supply time by increasing the number of soil cells in Earth-Battery, as shown in the 3D model of the green building infrastructure on the rooftop of a building. The main concern of the applied automated switching system is to make the hybrid system work more efficiently both in day and night time. The hybrid energy model may be affordable to needy people, produces no pollution, and also provides a sustainable alternative solution for all.

Acknowledgments First and foremost, the authors wanted to express their gratitude to the Almighty Allah and then their family members for their unconditional support. The authors would like to thank the Vice Chancellor of AIUB, Dr. Carmen Z. Lamagna Mam. The first author is very grateful to the Pro Vice Chancellor of AIUB and also this project supervisor, Professor Dr. Md. Abdur Rahman Sir, for his continued guidance for supervising this project to the successful completion with a positive impact on environment and come up as an alternative renewable energy solution. The article from Research on Sustainable Earth-Battery was linked to the **UN's SDGs 3** and **7** and contributing to overcoming some of the biggest problems in the world. This project also received the Gold Award in an international researcher's competition. The "2nd International Competition for Young Researchers 2022" organized by "UniV" association with international research collaboration community awarded this project as the "**Gold Award**." The newly introduced renewable energy project "Earth-Battery" project received more than 20 both of National and Divisional Awards from Bangladesh. Honorable thanks to the Power Cell, Power Division, Ministry of Power, Energy and Mineral Resources (MoPEMR)-Bangladesh for awarding the research project as a part of the BICCHURON-19 National Awarded Project of Power Cell based on renewable energy. Special thanks to Young-Bangla for patronizing the project and Green Delta Insurance Company Limited (GDIC) for providing award grant to this research project. Also, special thanks to Pathor Mukherjee for his outstanding work in visualizing the author's idea and Md. Sayeduzzaman, Md. Monir Hossain, and Firoz Ebne Jobaier for their technical support towards the project completion.

References

1. Borno, M.S.I., et al.: An empirical analysis of Sustainable Earth-Battery. Energy Rep. **7**, 144–151 (2021). ISSN 2352-4847, https://doi.org/10.1016/j.egyr.2021.06.026, https://www.sciencedirect.com/science/article/pii/S2352484721003917
2. Jing, W., et al.: Battery lifetime enhancement via smart hybrid energy storage plug-in module in standalone photovoltaic power system. J. Energy Storage. **21**, 586–598 (2019)
3. Al Badwawi, R., Abusara, M., Mallick, T.: A review of hybrid solar PV and wind energy system. Smart Sci. **3**(3), 127–138 (2015)
4. Wadi, M., et al.: Smart hybrid wind-solar street lighting system fuzzy based approach: case study Istanbul-Turkey. In: 2018 6th International Istanbul Smart Grids and Cities Congress and Fair (ICSG). IEEE (2018)
5. Meskani, A., Haddi, A.: Modeling and simulation of an intelligent hybrid energy source based on solar energy and battery. Energy Procedia. **162**, 97–106 (2019)
6. Deshmukh, M.K., Deshmukh, S.S.: Modeling of hybrid renewable energy systems. Renew. Sust. Energ. Rev. **12**(1), 235–249 (2008)
7. Nema, P., Nema, R.K., Rangnekar, S.: A current and future state of art development of hybrid energy system using wind and PV-solar: a review. Renew. Sust. Energ. Rev. **13**(8), 2096–2103 (2009)
8. Khalil, M.I., et al.: Hybrid smart grid with sustainable energy efficient resources for smart cities. Sust. Energy Technol. Assess. **46**, 101211 (2021)
9. Sayeduzzaman, M., Samiul Islam Borno, M., Yeasmin Fariya, K., Tamim Ahmed Khan, M.: A Method and Design for Developing an IoT-Based Auto-Sanitization System Powered by Sustainable Earth-Battery. In: Hossain, M.S., Majumder, S.P., Siddique, N., Hossain, M.S. (eds) The Fourth Industrial Revolution and Beyond. Lecture Notes in Electrical Engineering, vol 980. Springer, Singapore (2023). https://doi.org/10.1007/978-981-19-8032-9_47

Chapter 3
Upcycling Trash into Cash Through Repurposing the Bisasar Road Landfill Site into a Solar PV and Energy Storage Site in eThekwini Municipality

Leshan Moodliar and Innocent E. Davidson

3.1 Introduction

3.1.1 General Overview

South Africa has on average an installed capacity of 52,028 MW of generation capacity [1]. However, the availability of that capacity has been erratic for several years. Without a stable generation fleet, the load requirements of the country cannot be met, resulting in the implementation of rotational load shedding. Figure 3.1a illustrates the generation capacity available for 9 months of 2022. Figure 3.1b shows that unplanned maintenance is the main reason for generation loss in the country. Furthermore, the rising trend in unexpected breakdowns further signals the deterioration of generation infrastructure.

With severe capacity shortages, the Department of Mineral Resources and Energy (DMRE) has waivered the need for licensing generators from the initial 1 MW capacity to 100 MW to stimulate private sector generation [3]. Municipalities have historically operated within the distribution sector; however, with load shedding set to continue in the short term, municipalities can now capitalize on the opportunity to contribute to the generation capacity of South Africa. Further, electricity prices have risen by 128% [4] over the past 10 years, and future trends seem likely to continue as Eskom has applied for a 38.1% increase for the 2023/2024 financial year [5]. Therefore, municipalities owning and operating renewable energy generation and storage can better manage electricity increases and reduce prices

L. Moodliar (✉)
Durban University of Technology, Durban, South Africa
e-mail: 22289297@dut4life.ac.za

I. E. Davidson
Cape Peninsula University of Technology, Cape Town, South Africa

X. Wang (ed.), *Future Energy*, Green Energy and Technology,
https://doi.org/10.1007/978-3-031-33906-6_3

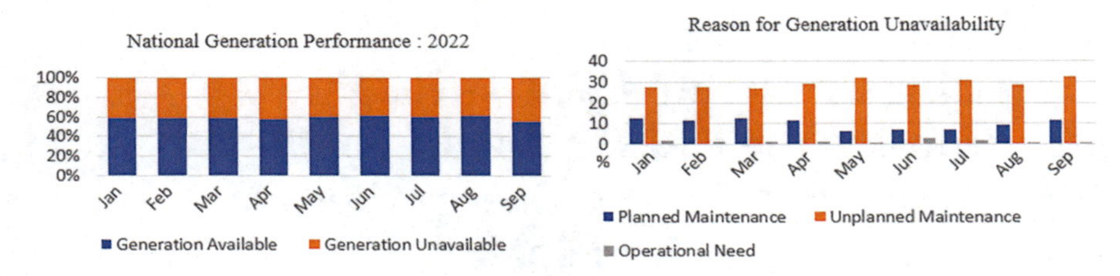

Fig. 3.1 (**a**) National generation performance. (Adapted from Ref. [2]). (**b**) Reasons for generation unavailability. (Adapted from Ref. [2])

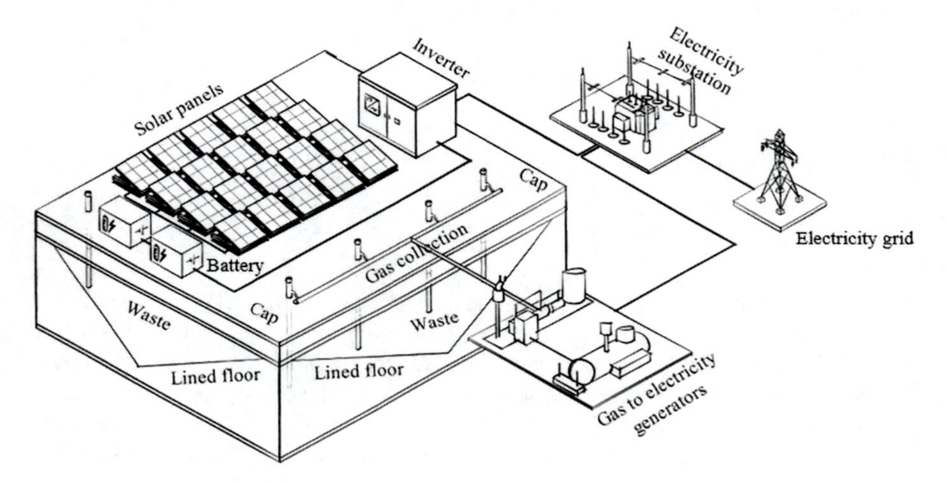

Fig. 3.2 Illustration of landfill with solar PV, battery storage, and landfill gas to electricity generation. (Adapted from Refs. [6, 7])

by leveraging renewable energy sources and energy arbitrage. Thus, the option to utilize solar energy and accompanying battery storage in landfills, as illustrated in Fig. 3.2, should not be disregarded. As a result, this study aims to evaluate the potential of repurposing the Bisasar landfill site within eThekwini Municipality into a solar PV and energy storage site.

3.1.2 Overview of the Landfill Site

Bisasar Road landfill is located northwest of Durban's city centre near Clare Estate. Since 1980, Durban Solid Waste (DSW) has managed the site. Before formal lining systems were established, 8000 metric tons of trash were dumped there between 1980 and 1996. At its peak, the landfill received 5200 tons of rubbish daily, making it one of the continent's busiest waste disposal sites [8–10]. Table 3.1 provides key facts pertaining to the Bisasar landfill site.

Table 3.1 Key facts about the Bisasar landfill site [11]

Year of establishment	Final landfill height (m)	Landfill size m^2	Capacity (m^3)	Average waste deposition rate per day (ton)
1980	122	440,000	21,000,000	3000

A range of containment cells allows for the ordered development of the landfill and the reintroduction of indigenous plants through the Plant Rescue UNIT (PRUNIT), enabling the site to align to its former condition [12]. In addition, a cap comprising layers of clay and sand is used to manage emissions from the landfill. Post-closure of the site, there is no prospect of the area being used for residential or general habitation [10]. The land would likely be continued to be used by the eThekwini Parks Department for plant staging as this process does not interfere with the post-closure requirements of the site [10]. With the site not explicitly earmarked for development use, installing solar PV and storage on the landfill presents an ideal opportunity. The supportive view of installing solar PV on landfill sites is also shared by many researchers, including [13–15].

3.2 Literature Review

Evaluating a landfill site for solar PV and battery storage installation requires careful consideration of the land suitability for solar PV harnessing. In addition, consideration must be given to land gradient and settlement. Other site considerations include the level of solar irradiation available, installation methods, and integration with the current generation systems [13, 14].

3.2.1 Land Gradient

Gentle slopes minimize solar design complexity and installation costs. As a result, gradients not exceeding 2–3% are advised for solar installations. Steep landfill slopes make foundational designs, wind loading calculations, and erosion management more complex, increasing expenses [6, 16]. To maximize electricity generation potential, fixed solar panels should face northward, tilted at an angle of 30° [17]. Gradients conforming to the tilt angle make for easier installation with optimum generation. Soil stability is also essential for solar PV designs and is carried out utilizing soil engineering and geotechnical data. Solar on slopes complicates soil stability analyses and is usually avoided in cases where flatter areas are available [6]. The contour lines and intervals for the landfill are shown in Fig. 3.3, where each contour line represents a 2 m rise in height. Evaluating the contour intervals, the variation in elevation at various parts of the landfill can be

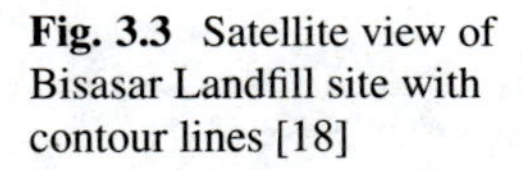

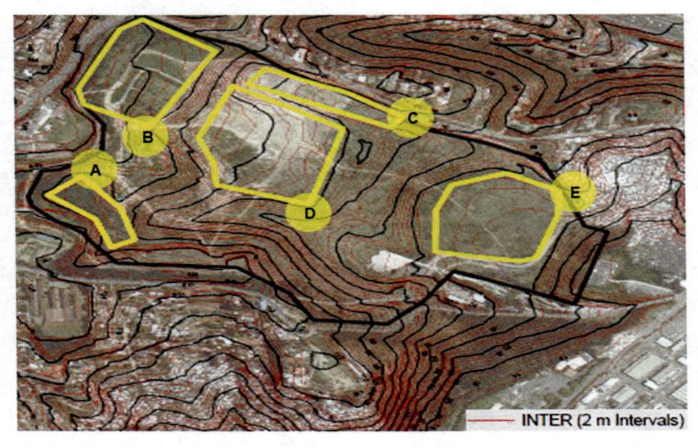

Fig. 3.3 Satellite view of Bisasar Landfill site with contour lines [18]

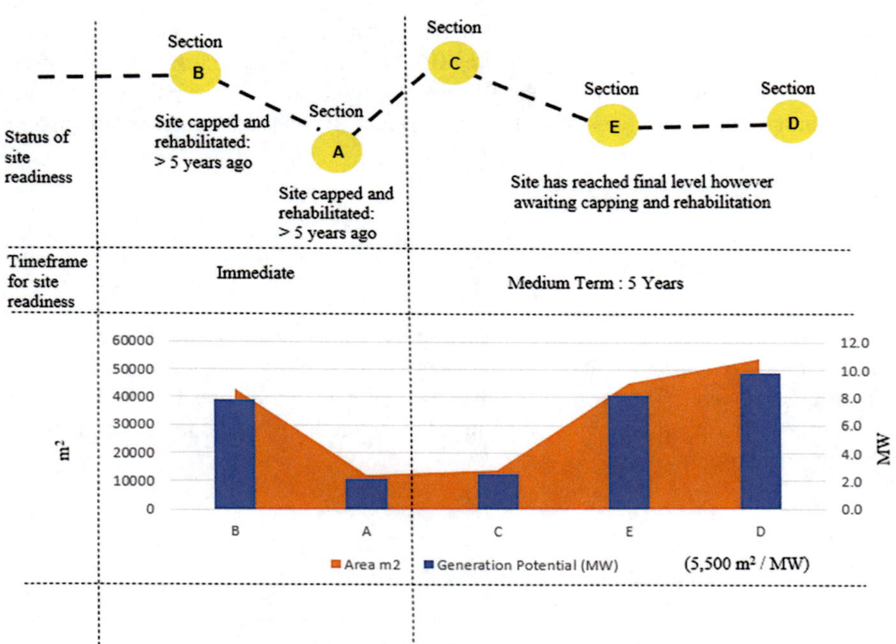

Fig. 3.4 Site readiness, area (m^2), and generational potential (MW) per section

understood. With due consideration of the need for flatter areas, five sections have been identified as plausible installation areas for solar PV and battery storage. The identified sections are illustrated in Fig. 3.4, together with the calculated areas and the generation potential.

3.2.2 Land Settlement

The Bisasar Road landfill was a dumping site for household waste, garden refuse, and rubble. Approximately 50% of the waste was biodegradable [8]. As a result of the mixed waste stream, the site would be subject to differential land settlement, where different parts of the site settle at different rates [19]. Recently closed areas are more suspectable to settlement [20]. Differential settlement places extreme stress on the solar PV mounting accessories and would eventually lead to the misalignment of racks and, ultimately, damage. With misalignment, solar PV production is reduced [14, 21]. Landfills capped fewer than 3 years ago should not be considered for solar installations due to rapid settlement rates [6]. Settlement rates should be monitored as actual settlement rates could vary across sites, depending on material and decomposition rates. The settlement could occur for up to 5 years post-closure of the landfill [22]. With waste heights known, settlement rates can be determined using the Power Creep Law, which describes the movement of viscous mediums under compression [23].

3.2.3 Solar Irradiation

The solar irradiation output is an essential parameter that determines the performance of the PV system. In addition to the irradiation data, rainfall, wind characteristics, and temperature are important parameters that influence performance. The PV and battery installation could also affect the flow of stormwater over the cap and change the levels of exposure to the sun and wind. These could affect cap integrity, leachate generation, vegetative cover, erosion management strategies, and evapotranspiration dynamics [6].

3.2.4 Solar Installation Methods

The simplest method for anchoring solar PV systems on landfills is using concrete slabs or footers. An advantage of this method is that there is no need to penetrate the cap [15]. However, these footings introduce weight to the surface and could affect the settlement rate. A common means to counteract this is by using adjustable solar racking and mounting systems that can be adjusted independently. Further, geo-grid reinforcement systems could also limit the impacts of differential settlement [6, 21].

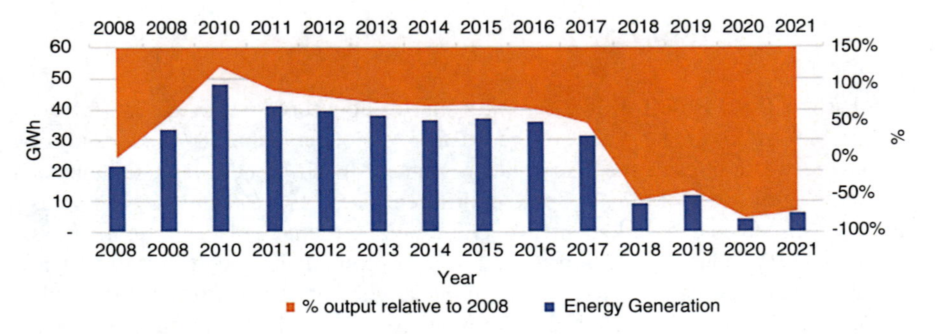

Fig. 3.5 Landfill gas to electricity generation analysis [9, 25]

3.2.5 Integration to Other Generation Sources

The Bisasar Road Landfill gas to electricity project was initially commissioned in 2008 and was the first landfill gas to electricity project in Africa [8], with an installation footprint of 8 MW; however, only 6.5 MW was installed. It generates between 30 and 40 GWh on average per annum after an initial capital investment of $6,253,585 [9, 24]. Gas is pumped from wells within the landfill into engines that drive generators that export electricity to the grid. However, after the landfill's closure, waste material was no longer deposited, resulting in a lower gas production and extraction yield. Consequently, the generator outputs have been significantly reduced, as shown in Fig. 3.5.

3.2.6 General Considerations

Landfill sites require periodic inspections and maintenance of the cap integrity, leachate systems, gas collection systems, and vegetation [6, 19]. Therefore, the PV system must be designed with these considerations as they may create practical layout problems. For example, the PV systems must be laid out to allow sufficient lawn mower access to maintain vegetation. In addition, solar PV and battery storage will attract incidents of theft and vandalism, and therefore, the site warrants additional measures of security to protect infrastructure [6]. Depending on the site circumstances, a combination of perimeter fencing, security alarms, cameras, lighting, and security guards would make for feasible mitigation solutions.

3.3 Research Methodology and Scenario Setting

As illustrated in Fig. 3.6, identifying the landfill and assessing its feasibility for installing solar PV and energy storage are the initial steps. The constraint criteria were established before modelling and optimization. The initial constraint was the space availability, while the investment criteria provided the second constraint. The available space was determined to be 55,000 m^2. However, only 50% was modelled to cater for any post-closure site requirements. The other constraint was the limited investment amount at the upper bound of $5,737,234 and a lower bound of $2,868,617. Finally, the design was optimized using the Hybrid Optimization Model for Multiple Energy Resources (HOMER) software tool. Project output feasibility indicators included the net present costs (NPC), the internal rate of return (IRR), the simple payback period (SPBP), and the levelized cost of energy (LCOE). The most optimized approach would maximize municipal revenue and carbon emission (CO_2) reductions while minimizing investment and space requirements.

Figure 3.7 shows the key input parameters used in the simulations. Costs for installation, replacement, and performance loss were set for the solar PV system. Parameters for battery storage were the type of battery, associated costs, and performance characteristics. In contrast, general modelling inputs included the expected project lifespan and relevant economic rates. Other factors that affected the project were the price of electricity and annual increases.

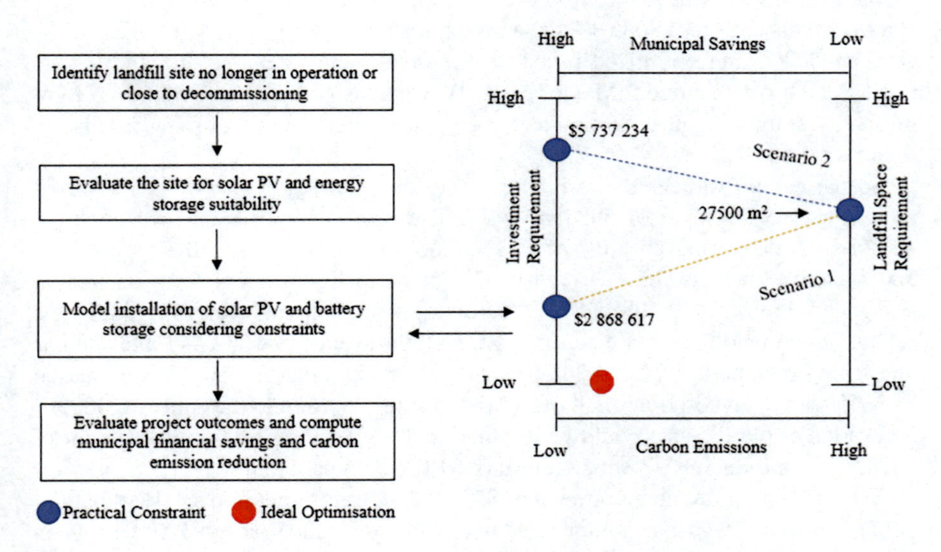

Fig. 3.6 Graphical representation of research methodology

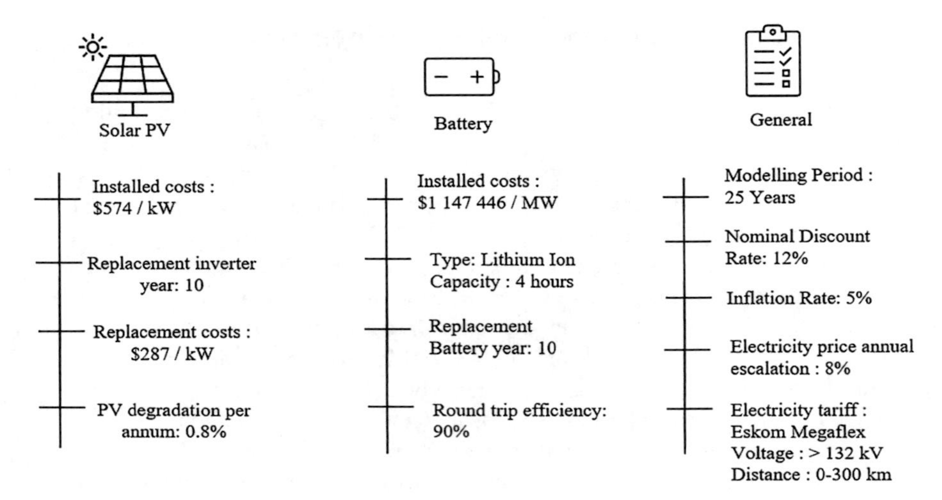

Fig. 3.7 Modelling inputs

3.4 Discussion and Summary of Results

To account for possible post-closure site requirements, only 50% of the 55,000 m^2 available area was evaluated for project development. The capital investment was set at a low threshold of $2,868,617 and a high threshold of $5,737,234. Two scenarios were modelled and optimized based on the investment thresholds for the lowest NPC. Scenario 1 featured 5 MW of solar PV, whereas scenario 2 comprised 5 MW of battery storage. Figure 3.8 provides a graphical representation of key results for scenarios 1 and 2.

Scenario 1 maximizes the available space for installing 5 MW solar PV based on a space requirement of 5500 m^2/MW. The project requires an investment of $2,868,617, delivering an NPC of $88,326,563 that produces an IRR of 24.20% and a simple payback of 4.79 years. The municipality benefits from an annual site generation of $519,798 with 5,225,786 kg of carbon emission (CO_2) savings. Including the landfill gas to electricity generation revenue of $582,443, the site has an earning potential of $1,102,241 per annum. The earning potential has increased by 89% with the incorporation of the 5 MW of solar PV. Should the entire 55,000 m^2 be available, the PV size could be doubled to 10 MW, requiring an investment of $5,737,234, producing a revenue return of $1,039,596 per annum.

Scenario 2 requires an investment of $5,737,234, translating into 5 MW of battery storage, which could be accommodated in an estimated land space of 1,800 m^2 when allowing for an installation footprint of 360 m^2 per MW installed [26]. The project delivers an NPC of $98,196,787 over 25 years and produces an IRR of 6.80%, with a payback of 16.47 years. The municipality benefits from an annual revenue generation of $361,654 by exploiting the principle of energy arbitrage, where the battery charges during off-peak periods and discharges during peak periods.

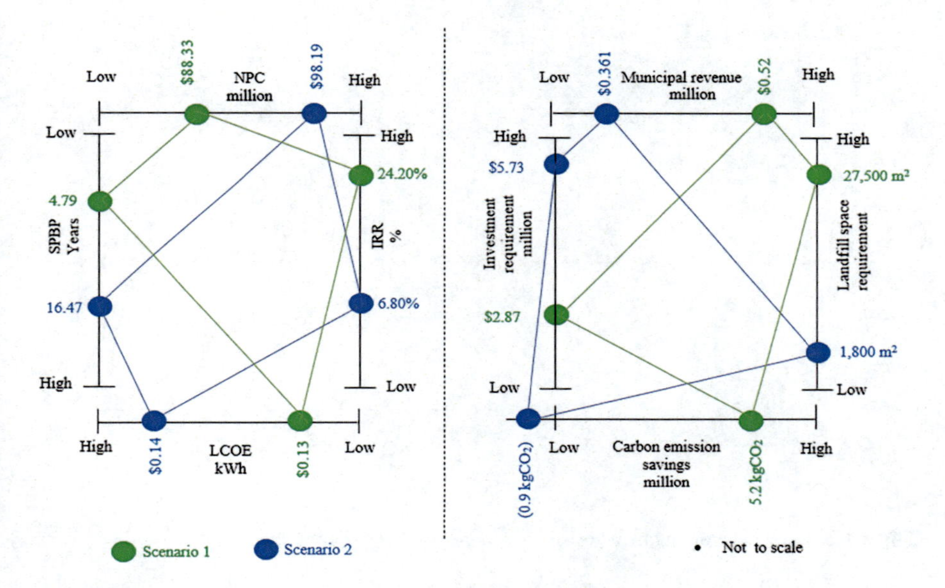

Fig. 3.8 Graphical representation of results for scenarios 1 and 2

Including the average landfill gas to electricity generation revenue of $582,443, the site has an earning potential of $944,097 per annum. The earning potential of the landfill site with the existing gas to electricity project has increased by 62% with the incorporation of the batteries. With more energy being generated and exported to the grid, the carbon emissions (CO_2) have also increased by 907,371 kg resulting in negative savings. Whilst this scenario is revenue generating, its IRR of 6.8% and a SPBP of 16.47 years is not attractive.

Combining scenarios 1 and 2 requires an investment of $8,605,852, with the project delivering an NPC of $88,133,047 that produces an IRR of 13.80% with a simple payback of 7.56 years. The municipality benefits from an annual revenue generation of $885,739 by harnessing solar energy and exploiting the principle of energy arbitrage, as shown in Fig. 3.9. The battery charges during off-peak periods and discharges during peak periods. Including the landfill gas to electricity generation revenue of $582,443, the site has an earning potential of $1,468,182 per annum.

3.5 Conclusion

Solar PV and batteries have demonstrated their ability to generate and store electricity. As technology advances, it is becoming more efficient and cost-effective. However, large-scale solar power generation and storage necessitate large tracts

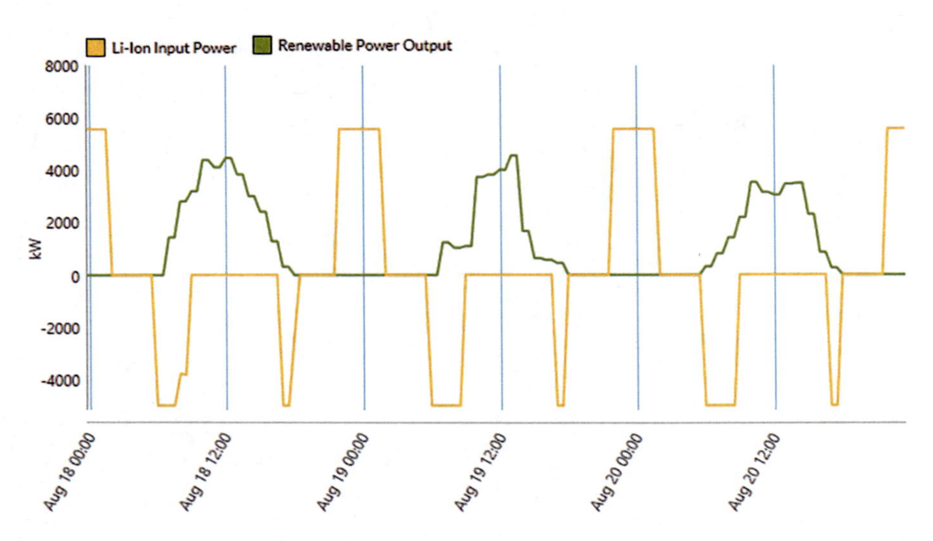

Fig. 3.9 Energy generation characteristics of scenarios 1 and 2 combined

of land, typically unavailable in metropolitan settings. Converting closed landfills into solar energy generating and storage facilities is a viable alternative. However, the landfill site conversion must consider geographical gradients, land settlements, solar irradiation, and current landfill gas to electricity-generating projects. In addition, particular consideration must be given to the placement of the solar system and mounting equipment, considering the landfill capping requirements, since destroying or breaching the cap might affect emission controls. The study utilized aerial photography and contour patterns of the landfill site to determine the areas appropriate for siting solar PV and battery storage. However, only two of the five identified portions are available for usage due to the completed capping and rehabilitation. The other three portions have reached their landfill limit however must be capped and rehabilitated. With only a minor section of the site still receiving garbage, the gas yields have decreased, resulting in a 70% decrease in the energy produced by the gas generators since their inception in 2008.

Using techno-economic modelling and the HOMER simulation tool, two scenarios were optimized in ascending investment order. With a low investment of $2,868,617, 5 MW of solar PV could be installed on 27,500 m^2, allowing for an annual revenue generation potential of $519,798 in addition to the landfill gas to electricity revenue generation of $582,443. Increasing the investment to $5,737,234 allowed for a project with 5 MW of battery storage utilizing only 1,800 m^2 and maximizing the principle of energy arbitrage, producing an annual revenue generation potential of $361,654 and $944,097 inclusive of the existing landfill gas to electricity project. Finally, combining the 5 MW of solar PV and 5 MW of battery storage at an investment of $8,605,852 would generate $1,468,182, including the existing landfill gas to electricity project.

Acknowledgement The authors thank eThekwini Municipality: Electricity Unit and the Cleansing & Solid Waste Unit for their assistance and data provision during the research project.

References

1. Ndlela, N.W., Davidson, I.E.: Reliability and security analysis of the Southern Africa Power Pool Regional Grid. In: IEEE PES/IAS Power Africa Conference, Rwanda, 22–26 Aug (2022). Available at: https://doi.org/10.1109/PowerAfrica53997.2022.9905389. Accessed 11 Sept 2022
2. Eskom: Outage performance. Available at: https://www.eskom.co.za/dataportal/outage-performance/ (2023). Accessed 05 Oct 2022
3. Department of Mineral Resources and Energy: Electricity Regulation Act, 2006: licensing exemption and registration notice. Available at: https://www.gov.za/sites/default/files/gcis_document/202104/44482gon374.pdf (2021). Accessed 05 Oct 2022
4. Eskom: Tariff history. Available at: https://www.eskom.co.za/distribution/tariffs-and-charges/tariff-history/ (2023). Accessed 01 Oct 2022
5. Eskom: Public hearing on the Eskom's fifth multiyear price determination (MYPD5) application for the 2022/23, 2023/24 and 2024/25 financial years. Available at: https://www.nersa.org.za/public-hearing-on-the-eskoms-fifth-multi-year-price-determination-mypd5-application-for-the-2022-23-2023-24-and-2024-25-financial-years/ (2022). Accessed 01 Oct 2022
6. The National Renewable Energy Laboratory (NREL) and US Environmental Protection Agency: Best practices for siting solar photovoltaics on municipal solid waste landfills. Available at: https://www.nrel.gov/docs/fy13osti/52615.pdf (2013). Accessed 05 Oct 2022
7. The Interstate Technology & Regulatory Council: Technical and Regulatory Guidance for Design, Installation, and Monitoring of Alternative Final Landfill Covers. ITRC, Washington, DC (2003)
8. Moodley, L.: Garden refuse composting as part of an integrated zero waste strategy for South African municipalities. MSc. University of Kwazulu-Natal. Available at: http://hdl.handle.net/10413/2857 (2010). Accessed 01 Oct 2022
9. Sewchurran, S., Davidson, I.E.: Optimisation and financial viability of landfill gas to electricity projects in South Africa. In: 2016 IEEE International Conference on Renewable Energy Research and Applications (ICRERA), Birmingham, UK, 20–23 Nov. Available at: https://doi.org/10.1109/ICRERA.2016.7884349 (2016). Accessed 12 Sept 2022
10. Wilson & Pass Inc: Bisasar Road Landfill Site Preliminary Remediation and Closure Plan. Durban Solid Waste Department, Durban (2015)
11. Couth, R., et al.: Delivery and viability of landfill gas CDM projects in Africa – a South African experience. Renew. Sust. Energy Rev. **15**(1), 392–403 (2011). https://doi.org/10.1016/j.rser.2010.08.004. Accessed 11 Sept 2022
12. Parkin, J., et al.: Extreme landfill engineering: developing and managing South Africa's busiest and largest landfill facilities. Available at: http://landfillconservancies.com/docs/parkin_et_al_extreme_landfill_engineering_paper_wastecon_2006.pdf (2006). Accessed 01 Oct 2022
13. Gu, G., et al.: Energetic and financial analysis of solar landfill project: a case study in Qingyuan. Int. J. Low-Carbon Technol. **17**, 214–221 (2022). Available at: https://doi.org/10.1093/ijlct/ctab095. Accessed 02 Oct 2022
14. Sangiorgio, S., Falconi, M.: Technical feasibility of a photovoltaic power plant on landfills. A case study. Energy Procedia **82**, 759–765 (2015). Available at: https://doi.org/10.1016/j.egypro.2015.11.807. Accessed 02 Oct 2022
15. Szabó, S., et al.: A methodology for maximizing the benefits of solar landfills on closed sites. Renew. Sust. Energy Rev. **76**, 1291–1300 (2017). Available at: https://doi.org/10.1016/j.rser.2017.03.117. Accessed 11 Sept 2022

16. Tansel, B., Varala, P.K., Londono, V.: Solar energy harvesting at closed landfills: energy yield and wind loads on solar panels on top and side slopes. Sust. Cities Soc. **8**, 42–47 (2013). Available at: https://doi.org/10.1016/j.scs.2013.01.004. Accessed 12 Sept 2022
17. Le Roux, W.G.: Optimum tilt and azimuth angles for fixed solar collectors in South Africa using measured data. Renew. Energy. **96**, 603–612 (2016)
18. EThekwini Municipality: Information Management Unit – GIS Department: Bisasar Road landfill site [Online]. EThekwini Municipality, Durban (2022)
19. Grace, M.: Landfill Turned into Solar-Powered Generation Facility, pp. 1–20. Landfill Technologies (2015). Available at: http://agruamerica.com/wp-content/uploads/2015/07/Landfill-Turned-Into-Solar-Powered_Final.pdf. Accessed 12 Sept 2022
20. Ong, S., et al.: Land-use requirements for solar power plants in the United States, No. NREL/TP-6A20-56290, National Renewable Energy Lab (NREL), Golden, CO United States. Available at: https://doi.org/10.2172/1086349 (2013). Accessed 05 Oct 2022
21. Sampson, G.: Solar Power Installations on Closed Landfills: Technical and Regulatory Considerations. Remediation and Technology Innovation, Washington, DC (2009). Available at: https://www.clu-in.org/download/studentpapers/Solar-Power-Installations-on-Closed-Landfills-Sampson.pdf. Accessed 05 Oct 2022
22. Munsell, D.R.: Closed landfills to solar energy power plants: estimating the solar potential of closed landfills in California. MSc. University of Southern California (2013)
23. Kumar, S.: Settlement prediction for municipal solid waste landfills using power creep law. Soil Sediment Contam. **9**(6), 579–592 (2000)
24. Pather-Elias, S., et al.: EThekwini Municipality project: Bisasar Road and Mariannhill landfill sites grid-tied electricity generation. Available at: www.cityenergy.org.za (n.d.). Accessed 02 Oct 2022
25. EThekwini Municipality: Electricity Unit: Bisasar Road Landfill Site Bills 2008 to 2021. EThekwini Municipality Electricity Unit, Durban (2022)
26. FM Global: Property loss prevention data sheets: electrical energy storage systems. Available at: https://www.fmglobal.com/research-and-resources/fm-global-data-sheets (2023). Accessed 05 Oct 2022

Chapter 4
Economic Analysis Approach for Energy Supply System Considering Load Characteristics

Wei Liu, Xiaoli Meng, Zhijie Yuan, Yuanhong Liu, Xihai Zhang, and Hong Liu

4.1 Introduction

Energy is one of the driving forces for the development of human society, promoting the progress of human society from an agricultural society to an industrial society [1]. However, in the face of enormous energy demand and carbon emission pressure, it is urgent to develop a feasible energy supply paradigm at the present stage, which combines clean energy, dominant energy, and load characteristics [2]. With the increase in energy coupling scenarios, the demand for electricity, gas, and heat (cooling) is increasingly common within energy buildings [3]. Therefore, the integrated energy system (IES) is regarded as a tailor-made scheme for modern society [4]. Furthermore, it has been applied in various scenarios with the aid of state-of-the-art artificial intelligence technology [5, 6]. However, considering the existence of multiple load characteristics, the cost-optimal energy supply system varied in different scenarios. Hence, it is of great importance to propose an economic analysis approach for regional energy supply system planning.

Since energy supply system planning is a potential solution to meet energy demand and carbon emissions, many studies have been implemented to resolve this dilemma by considering the reliability of energy supply equipment [7–9], IES comprehensive evaluation [10], hybrid energy storage system [11], energy stations-pipe networks synergy planning [12], and peer-to-peer energy trading [13]. However, the aforementioned literature concentrates more on energy supply equipment planning and operation, but the energy supply paradigm is not highly explored. The load

W. Liu · X. Meng · Y. Liu
China Electric Power Research Institute, Beijing, P.R. China
e-mail: liuwei75@epri.sgcc.com.cn; mengxl@epri.sgcc.com.cn; liuyuanhong@epri.sgcc.com.cn

Z. Yuan · X. Zhang (✉) · H. Liu
School of Electrical and Information Engineering, Tianjin University, Tianjin, P.R. China
e-mail: zhijieyuan@tju.edu.cn; xihaizhang@tju.edu.cn; liuhong@tju.edu.cn

X. Wang (ed.), *Future Energy*, Green Energy and Technology,
https://doi.org/10.1007/978-3-031-33906-6_4

within the target region may consist of industrial load, commercial load, office load, or residential load [14]. The characteristics of these loads are different, i.e., the energy usage of commercial load and office load are in the daytime during the work day, but the peak time of residential load is in the evening. The industrial load prefers to start at night due to the time-of-use price. Hence, it is significant to propose an economic evaluation strategy in which a suitable energy supply system is given based on the characteristics of the load.

In this chapter, an economic analysis approach for regional energy supply systems is proposed. Firstly, three mainstream energy supply systems are given based on the dominant energy imported types, i.e., electricity-based, gas-based, or electricity–gas joint-based. Secondly, an economic analysis approach is implemented to estimate the cost-optimal energy supply system. The economic analysis approach is based on the life cycle cost (LCC) model with the comprehensive consideration of initial investment cost, annual exogenous cost, and salvage cost. Finally, the LCC model is applied in a real regional IES case to analyze the cost-optimal energy supply system scheme. It is worth pointing out that industrial loads are usually supplied separately by their own energy stations and are out of the scope of this chapter. Furthermore, the modern building integrates commercial and office properties together, and we use novel commercial loads to substitute traditional commercial loads and office loads in this study.

The remainder of this chapter is organized as follows. Section 4.2 states three mainstream energy supply systems considering the different energy-dominant types. Following, the economic evaluation strategy of the energy supply system is introduced in Sect. 4.3. Then, the case study is shown in Sect. 4.4. At last, conclusions are drawn in Sect. 4.5.

4.2 The Energy Supply Systems

Electricity, gas, and heat are the most common energy on the user side. Hence, we investigate three mainstream energy supply systems, i.e., electricity-based energy supply systems, gas-based energy supply systems, and electricity–gas joint-based energy supply systems [15]. The energy supply equipment and energy coupling methods will be described in detail in the following. It is worth noting that the abovementioned energy supply system can couple with other types of energy, such as hydrogen energy [16]. However, considering that the usage of hydrogen is not universal, we just mixed the hydrogen with natural gas and took it as a special case of the gas-based energy supply system.

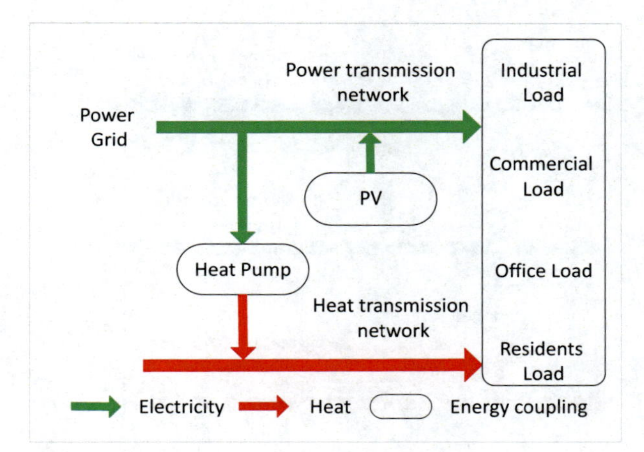

Fig. 4.1 Electricity-based energy supply system

4.2.1 Electricity-Based Energy Supply System

The electricity-based energy supply system is the most common energy supply system among the residents, which is shown in Fig. 4.1.

Within the electricity-based energy supply system, the power energy is imported by distributed generator or the power grid. The heat energy is from energy coupling units, such as heat pumps. It is worth pointing out that there are three types of heat pumps, i.e., water-source heat pumps, air-source heat pumps, and ground-source heat pumps. The applicability of these heat pumps is limited by the geographical structure or regional restrictions [17]. Furthermore, the storage unit is an optional unit for electricity and heat, which is up to the demand response potential of the target energy supply region.

4.2.2 Gas-Based Energy Supply System

The gas-based energy supply system is a universal energy supply system, especially in industrial scenarios or commercial scenarios. The diagram of the gas-based energy supply system is shown in Fig. 4.2.

The gas-based energy system has more energy imported types, and hence it is more complex than the electricity-based energy system. The gas boiler is a tailor-made energy coupling unit with more economic advantages. Furthermore, the gas boiler can provide both domestic water and heat demand.

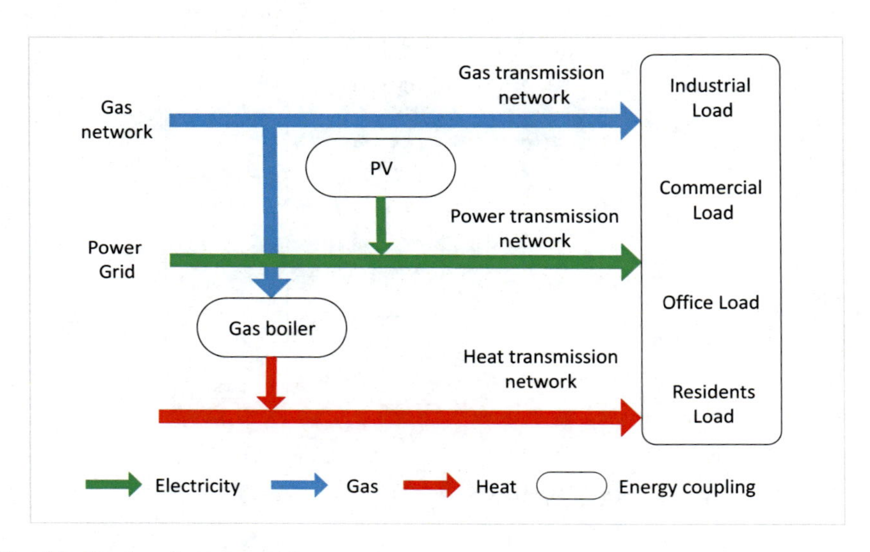

Fig. 4.2 Gas-based energy supply system

4.2.3 Electricity–gas Joint-Based Energy Supply System

The electricity–gas joint-based energy supply system is an emerging energy supply system, and the efficiency is further improved by the interaction of the electrical-gas-heat supply subsystem. The electricity–gas joint-based energy supply system is shown in Fig. 4.3.

Different from the abovementioned energy supply system, the co-generation, combined heat and power (CHP) unit consists of a power plant, generator, waste heat boiler, and heat exchanger as shown in Fig. 4.4. The power plant within the CHP unit is a gas turbine or internal combustion engine. The power plant utilizes natural gas to obtain high-temperature flue gas. Then, the generator is driven and supplies electric power. Meanwhile, the waste heat is supplied to the end-user by the waste heat boiler and heat exchanger. The CHP can work in two modes, i.e., power determined by heat or heat determined by power. Therefore, the heat–electricity ratio is essential for calculating output electrical/heat power. Furthermore, considering the multi-step power price or time-of-use power price scheme, the CHP can generate electricity for self-usage or sell back electricity to the power grid to reduce the local operation cost. Moreover, the heat pump within the electricity–gas joint-based energy supply system is an auxiliary electricity–heat coupling unit, which can further improve energy utilization efficiency.

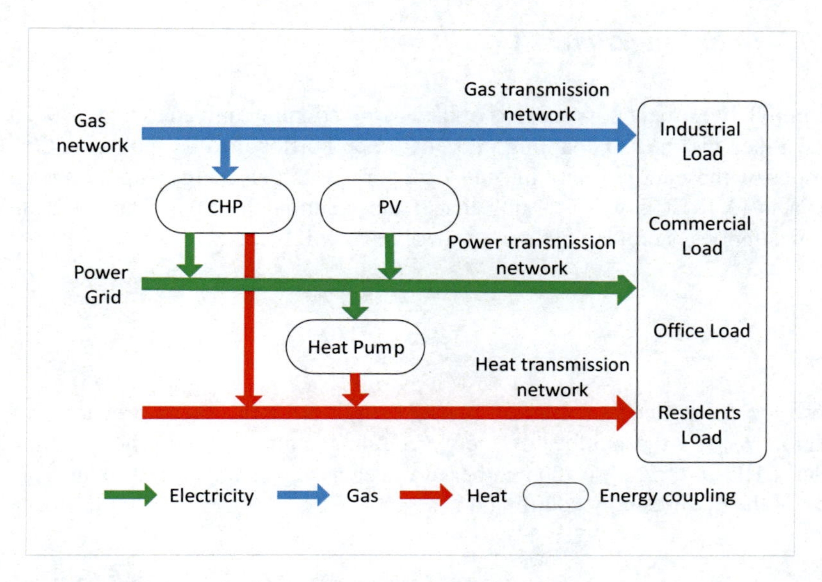

Fig. 4.3 Electricity–gas joint-based energy supply system

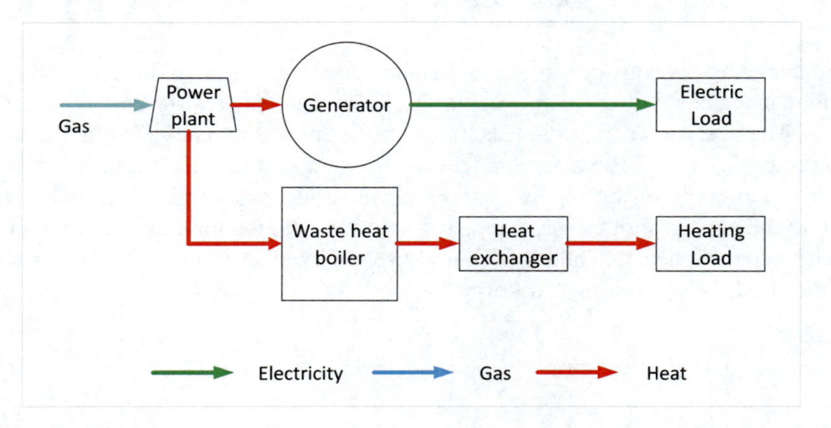

Fig. 4.4 Principle diagram of CHP unit

4.3 Economic Evaluation Strategy of Energy Supply System

The energy supply system for a target region is up to the economic prospect within the life cycle. Therefore, the LCC model is proposed to evaluate the economic potential of different energy supply systems in this study [18]. The LCC model consists of three economic metrics, i.e., initial investment cost, annual exogenous cost, and salvage cost. The mathematical model will be described in detail in the following.

4.3.1 Initial Investment Cost

The initial investment cost consists of the energy coupling unit costs and installation costs. Since the energy coupling unit cost is proportional to the unit quantity, the formula of the energy coupling unit cost for the target energy supply system is shown in (4.1). It is worth noting that if the mentioned units do not exist in the energy supply system, the number of these units will be zero.

$$C_1 = \sum_{i=1}^{N_{unit}} K_i m_i \tag{4.1}$$

where the C_1 is the total cost of energy coupling for the specific energy supply system. N_{unit} is the number of energy coupling units, including heat pump, gas boiler, CHP, and PV. K_i is the number of target energy coupling units, and m_i is the price of the target energy coupling unit.

4.3.2 Annual Exogenous Cost

The annual exogenous cost includes maintenance, operation, and carbon tax. The maintenance cost is the fee for regular/irregular maintenance for energy coupling units. The operation cost is the fees for electrical and gas purchased from the power grid or gas network. The electrical fees are the sum cost of the electricity imported from the power grid or exported to the power grid. It is worth pointing out that due to the existence of energy couples within the IES, the total electricity demand includes part of heat demand, which is converted by electrical energy. The formula of electrical fees is shown as follows:

$$C_e = \left[L_e - Q_{\text{PV}} - Q_{\text{CHP}} + \frac{Q_{eh}}{\varphi_{eh}}\right]^+ \pi_e^b + \left[L_e - Q_{\text{PV}} - Q_{\text{CHP}} + \frac{Q_{eh}}{\varphi_{eh}}\right]^- \pi_e^s \tag{4.2}$$

where C_e is the total electrical fees. L_e is the volume of electricity used on the user side. Q_{PV} is the volume of electricity generated by PV. Q_{CHP} is the volume of electricity converted from gas by CHP. Q_{eh} is the heat converted from the heat pump and φ_{eh} is heat pump efficiency. π_e^b and π_e^s are the unit price of the electricity imported or exported from the power grid, respectively.

The gas fees are evaluated by the usage of gas. Like electricity, it also exists the phenomenon of energy couples. The formula of gas fees is shown as follows:

$$C_g = \left(\frac{Q_{\text{CHP}}}{q\varphi_{\text{CHP}}} + \frac{Q_{gh}}{q\varphi_{gh}}\right)\pi_g \tag{4.3}$$

where C_g is the total gas fee. φ_{CHP} is the gas conversion efficiency of CHP. φ_{gh} is the gas conversion efficiency of the gas boiler. q is the calorific value of gas, which is set as 35 MJ/Nm^3. π_g is the price of unit gas.

The heat demand can be met by a heat pump, gas boiler, or CHP in the different energy systems, and the equilibrium formula is shown as follows:

$$L_h = Q_{\mathrm{CHP}} R \varphi_h + Q_{gh} + Q_{eh} \tag{4.4}$$

where the L_h is the volume of heat demand on the user side, R is the heat-electricity ratio of the CHP unit, and φ_h is the heat efficiency of the CHP unit.

Considering maintenance fees are proportional to the volume of electric, gas, and heat exported by energy coupling units, the formula is shown as follows:

$$C_{ma} = \sum_{i=1}^{N_{unit}} Q_i \sigma_i \tag{4.5}$$

where Q_i is the real output power of the corresponding energy coupling units, and σ_i is the unit maintenance cost for the energy coupling units.

The carbon tax is imposed on carbon dioxide emissions into the air. Since the release of the “dual carbon” policy, carbon emissions have been a big concern in selecting the energy supply system. The carbon tax is evaluated by

$$C_{carbon} = \rho(\tau_e Q_e + \tau_g Q_g) \tag{4.6}$$

where Q_e and Q_g are the volumes of electricity and gas imported from the power grid and gas network. τ_e and τ_g are the equivalent emission factor for the power grid and gas network, respectively. ρ is a unit equivalent carbon tax.

The total annual exogenous cost is the sum of operation cost and maintenance cost, and the formula is shown as follows:

$$C_2 = \sum_t C_{e,t} + C_{g,t} + C_{ma,t} + C_{carbon,t} \tag{4.7}$$

where t is the time slot within the service life.

4.3.3 Salvage Cost

The salvage cost is the depreciation charge of the energy coupling unit within the energy supply system, which is calculated as a fixed percentage of the original investment cost. The formula is shown as follows:

$$C_3 = 5\% C_{unit} \tag{4.8}$$

4.3.4 Life Cycle Cost Model

Based on the mathematical model of initial investment cost, annual exogenous cost, and salvage cost, considering the service life and discount rate, the LCC model is described using

$$C = C_1 \frac{r(1+r)^N}{(1+r)^N - 1} + C_2 - C_3 \frac{r}{(1+r)^N - 1} \tag{4.9}$$

where C is the equivalent annual cost within the life cycle, C_1 is the initial investment cost, C_2 is the annual exogenous cost, C_3 is salvage cost, N is service life and set as 25 years, and r is the discount rate, which is considered as 7%.

The overall economic evaluation strategy of the energy supply system is to minimize the low cost under operational constraints. The final optimization model is mixed-integer programming, shown as follows:

$$\min \quad C \tag{4.10a}$$

$$s.t. \quad (4.1)\text{–}(4.9) \tag{4.10b}$$

$$0 \leq Q_i \leq K_i P_i \tag{4.10c}$$

$$Q_e = L_e - Q_{\mathrm{PV}} - Q_{ge} + \frac{Q_{eh}}{\varphi_{eh}} \tag{4.10d}$$

$$Q_g = \frac{Q_{\mathrm{CHP}}}{q\varphi_{\mathrm{CHP}}} + \frac{Q_{gh}}{q\varphi_{gh}} \tag{4.10e}$$

where the integer decision variables (K_i) mean the number of target energy coupling units mentioned in Sect. 4.2, and the continuous decision variables (Q_i) mean the real output power of energy coupling units (4.10c) denotes that the maximum output power of the target energy coupling unit, P_i, is the rated power of energy coupling units (4.10d) and (4.10e) are the volume of electricity and gas imported/exported to the power grid or gas network.

4.4 Case Study

The numerical results of the economic evaluation are presented in this section. The formulation is programmed in Python, and the solver is Gurobi runs on a Windows 10 PC with 64 GB of RAM, 14 logical cores, and a speed of 3.3 GHz. The planning area is a residential–commercial mixed area in China.

4.4.1 Basic Data

Considering the climate and building environment in China, most of the heat demand is in the winter, and the cold load is in the summer which is realized by the air conditioners. Therefore, the cooling load is usually added to the electrical load after being equivalent. Supposing that the behavior of energy consumption in each month is similar, a typical heating/power day in each month is selected to simulate the whole month. The typical daily load is shown in Fig. 4.5.

The time-of-use residential and commercial electricity price is summarized as follows. The peak time price (8:00–10:00 and 18:00–22:00) is 0.5583 CNY/kWh and 1.234 CNY/kWh, respectively. The valley price (1:00–6:00 and 23:00–24:00) is 0.3583 CNY/kWh and 0.5625 CNY/kWh, respectively. The electricity price for rest time is 0.5283 CNY/kWh and 0.7625 CNY/kWh. The feed-in-back electricity price is 0.06 CNY/kWh. The gas price for residential and commercial is 2.6 CNY/m^3 and 5 CNY/m^3, respectively. The equivalent emission factor of the power grid and gas

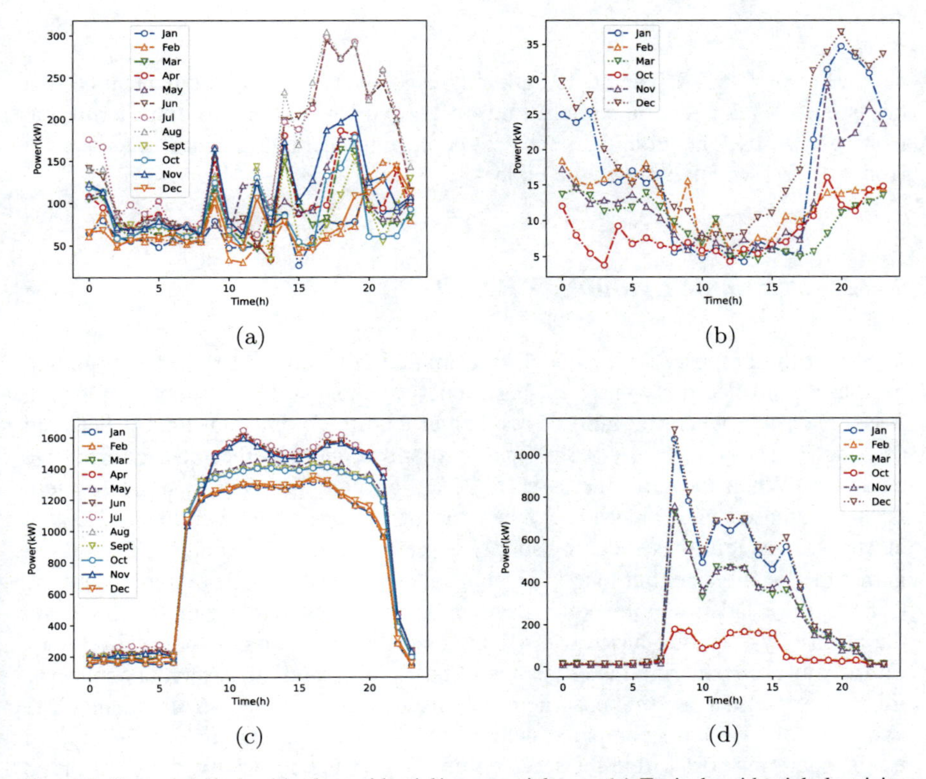

Fig. 4.5 Typical daily load in the residential/commercial area. (a) Typical residential electricity load. (b) Typical residential heat load. (c) Typical commercial electricity load. (d) Typical commercial heat load

Table 4.1 Energy supply system equipment and parameters

Parameters	Value
Rated power of CHP P_{CHP}/kW	300
CHP price m_{CHP} / (k · CNY)	9670
CHP unit maintenance cost σ_{CHP} /[CNY · kW^{-1}]	0.025
Heat-electricity ratio R	1.6
CHP heat efficiency φ_h	42%
CHP gas conversion efficiency φ_{CHP}	60%
Rated power of gas boiler P_{gh}/kW	300
Gas boiler price m_{gh} / (k · CNY)	96
Gas boiler unit maintenance cost σ_{gh} /[CNY · kW^{-1}]	0.05
Gas boiler conversion efficiency φ_{gh}	90%
Rated power of heat pump P_{hp}/kW	300
Heat pump price m_{hp} / (k · CNY)	125
Heat pump unit maintenance cost σ_{hp} /[CNY · kW^{-1}]	0.097
Heat pump efficiency φ_{hp}	440%
Rated power of PV P_{pv}/kW	35
PV price m_{pv} / (k · CNY)	300

network is 0.80 kg/kWh and 0.19 kg/kWh, respectively. The unit equivalent carbon tax is 0.3 CNY/kg. The maximum limit of PV panels is set as 10 considering the hosting capacity. The ground-source heat pump and other equipment parameters are from [19–21], which is shown in Table 4.1.

4.4.2 Economic Evaluation Result

Based on the proposed LCC model, the simulation result of the electricity-based, gas-based, and electricity–gas joint-based energy supply system is shown in Fig. 4.6. It is shown that when the ratio of residential load to commercial load is between 0% and 7%, the electricity-based energy supply system is the most competitive paradigm. When the ratio increases to 12%, the potential energy supply system is the gas-based energy system. When the percentage is greater than 12%, the electricity–gas joint-based energy supply is preferred in the target area. The reason is that the local residential load is much smaller than the commercial load, and the CHP unit is relatively more expensive than other energy coupling units. Therefore, the electricity–gas joint-based energy supply system is not a cost-optimal scheme. Furthermore, considering the efficiency of heat pumps and the carbon tax concerns for end-users, the operating cost increases sustainably with the rise of demand. The extra cost of CHP units compared with other energy coupling units will be covered by the operating cost within the service time. Hence, the electricity–gas joint-based energy supply system is chosen in these scenarios. It is worth pointing out that the electricity–gas joint-based energy supply in the 100% commercial load scenario can

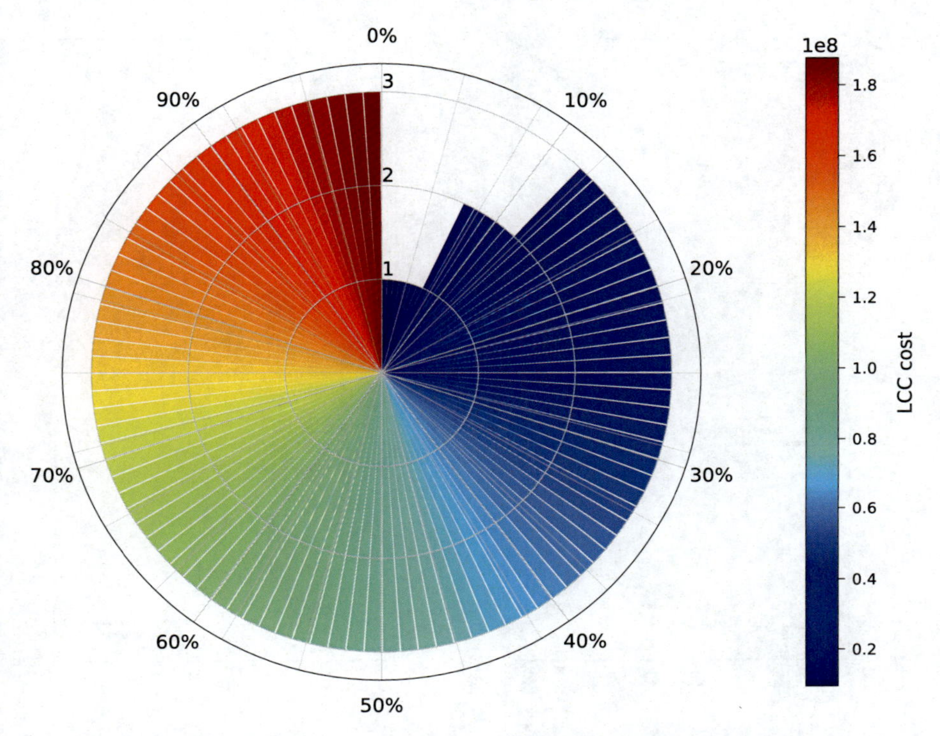

Fig. 4.6 Economic evaluation of three energy supply systems (1: electricity-based, 2: gas-based, and 3: electricity–gas joint-based)

save 18.68% and 17.50% when compared with the electricity-based energy supply system and gas-based energy supply system, respectively.

4.4.3 Sensitivity Analysis

Considering the policy of energy prices is different in different regions, the sensitivity analysis of energy price is carried out, and the result is shown in Fig. 4.7. The scale factor is the ratio between the spot price and the baseline price mentioned in this chapter.

As shown in Fig. 4.7a and b, when encountering the low local electricity price and 100% residents load case, the electricity-based or gas-based energy supply system is preferred by the end-user. With the increase in local electricity prices, the electricity–gas joint-based energy supply system is cost-optimal since the CHP unit can generate more power at the peak time to mitigate electricity costs. For the 100% commercial load scenario, the advantage of the CHP unit is more apparent when compared with the other two energy supply systems. It is worth mentioning

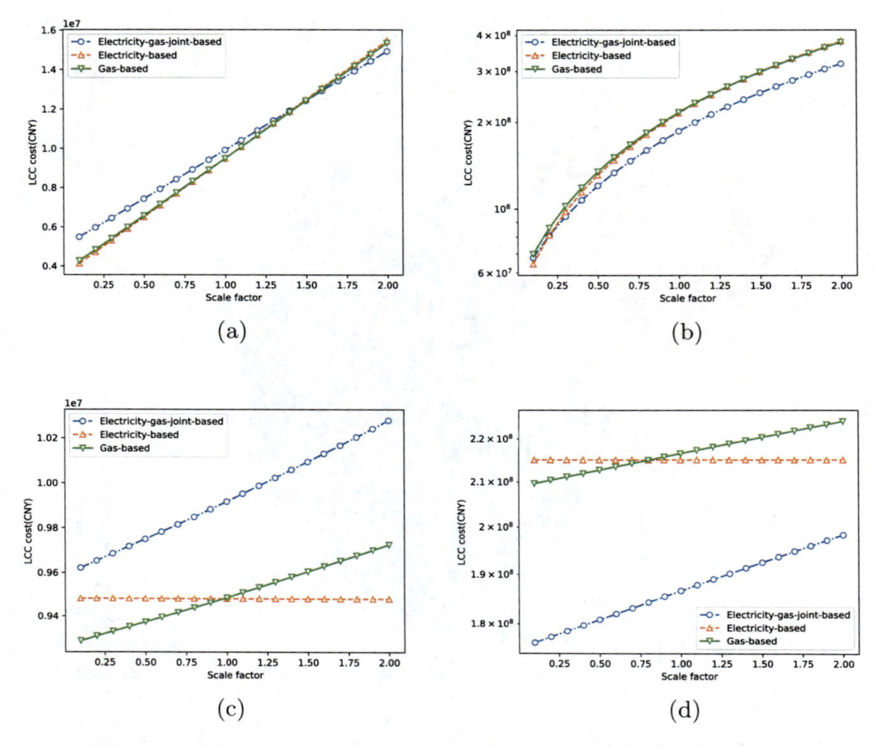

Fig. 4.7 Electricity/gas price sensitivity analysis. (**a**) Electricity price to residential load. (**b**) Electricity price to commercial load. (**c**) Gas price to residential load. (**d**) Gas price to commercial load

that since the lack of energy coupling units can convert to electricity, the electricity price is not sensitive to the electricity-based or gas-based energy supply system compared with the electricity–gas joint-based energy supply system.

Considering that the electricity-based energy supply system is immune to the policy of gas price, the LCC cost keeps the same as shown in Fig. 4.7c and d. In the 100% residential load case, due to the high cost of CHP units, the electricity–gas joint-based energy supply system is not a good choice in the view of the economy. The gas-based energy supply system is the cheapest when the scale factor is less than 0.9. However, when the scale factor is greater than 0.9, the electricity-based energy supply system is more promising. In the 100% commercial load scenario, the CHP unit can make full use of electricity and gas. It is always the most economical energy supply system when compared with the other two energy supply systems.

4.5 Conclusions

This chapter proposes an LCC-based economic analysis approach for the energy supply system. Through analyzing dominant energy imported, the energy supply system is classed into three types. Then, the LCC model is implemented to evaluate the economic prospect of the different energy supply systems. The simulation result shows that the cost-optimal energy supply system varied from the changes in load characteristics. The electricity–gas joint-based energy supply system prefers a high percentage of commercial load scenarios. The electricity-based energy supply system is more inclined to the high rate of residents' load scenarios. The applicable gas-based energy supply system scenario lies between the scenarios mentioned above.

Acknowledgments This work is supported by the Science and Technology Project of State Grid (5400-202155371A-0-0-00).

References

1. Gao, L., Hwang, Y., Cao, T.: An overview of optimization technologies applied in combined cooling, heating and power systems. Renew. Sust. Energ. Rev. **114**, 109344 (2019). https://doi.org/10.1016/j.rser.2019.109344
2. Zhou, Y., Manea, A.N., Hua, W., Wu, J., Zhou, W., Yu, J., Rahman, S.: Application of distributed ledger technology in distribution networks. Proc. IEEE **110**(12), 1963–1975 (2022). https://doi.org/10.1109/JPROC.2022.3181528
3. Belussi, L., Barozzi, B., Bellazzi, A., Danza, L., Devitofrancesco, A., Fanciulli, C., Ghellere, M., Guazzi, G., Meroni, I., Salamone, F., Scamoni, F., Scrosati, C.: A review of performance of zero energy buildings and energy efficiency solutions. Journal of Building Engineering **25**, 100772 (2019). https://doi.org/10.1016/j.jobe.2019.100772
4. Feng, C., Liao, X.: An overview of "energy + internet" in China. J. Clean. Prod. **258**, 120630 (2020). https://doi.org/10.1016/j.jclepro.2020.120630
5. Chen, Y., Chen, Y., Zhang, G., Paul, P., Wu, T., Zhang, X., Rong, H., Ma, X.: A survey of learning spiking neural p systems and a novel instance. Int. J. Unconv. Comput. **16**(2–3), 173–200 (2021)
6. Zhang, G., Zhang, X., Rong, H., Paul, P., Zhu, M., Neri, F., Ong, Y.S.: A layered spiking neural system for classification problems. Int. J. Neural Syst. **32**(08), 2250023 (2022). https://doi.org/10.1142/S012906572250023X
7. Zhang, X., Zhang, G., Paul, P., Zhang, J., Wu, T., Fan, S., Xiong, X.: Dissolved gas analysis for transformer fault based on learning spiking neural p system with belief AdaBoost. Int. J. Unconv. Comput. **16**(2–3), 239–258 (2021)
8. Wang, J., Zhang, X., Zhang, F., Wan, J., Kou, L., Ke, W.: Review on evolution of intelligent algorithms for transformer condition assessment. Frontiers in Energy Research **10** (2022). https://doi.org/10.3389/fenrg.2022.904109
9. Wang, J., Zhang, X., Li, P., Liu, X.: A comprehensive survey on transformer fault diagnosis and operating condition prediction. In: 2020 IEEE 4th Conference on Energy Internet and Energy System Integration (EI2), pp. 579–584 (2020). https://doi.org/10.1109/EI250167.2020.9346637

10. Liu, Y., Liu, W., Meng, X., Zhang, X., Ge, S., Liu, H.: Comprehensive evaluation of regional integrated energy system using the fuzzy analytic hierarchy process based on set pair analysis. In: 2022 IEEE/IAS Industrial and Commercial Power System Asia (I&CPS Asia), pp. 2019–2024 (2022). https://doi.org/10.1109/ICPSAsia55496.2022.9949940
11. Wang, Y., Zhang, Y., Xue, L., Liu, C., Song, F., Sun, Y., Liu, Y., Che, B.: Research on planning optimization of integrated energy system based on the differential features of hybrid energy storage system. Journal of Energy Storage **55**, 105368 (2022). https://doi.org/10.1016/j.est.2022.105368
12. Liu, H., Xiang, C., Ge, S., Zhang, P., Zheng, N., Li, J.: Synergy planning for integrated energy stations and pipe networks based on station network interactions. Int. J. Electr. Power Energy Syst. **125**, 106523 (2021). https://doi.org/10.1016/j.ijepes.2020.106523
13. Zhang, X., Ge, S., Liu, H., Zhou, Y., He, X., Xu, Z.: Distributionally robust optimization for peer-to-peer energy trading considering data-driven ambiguity sets. Appl. Energy **331**, 120436 (2023). https://doi.org/10.1016/j.apenergy.2022.120436
14. Pansini, A.J.: Chapter 2—load characteristics. In: Electrical Distribution Engineering, 3rd edn., pp. 37–50. River Publishers, New York (2020)
15. Feng, L., Mears, L., Beaufort, C., Schulte, J.: Energy, economy, and environment analysis and optimization on manufacturing plant energy supply system. Energy Convers. Manag. **117**, 454–465 (2016). https://doi.org/10.1016/j.enconman.2016.03.031
16. Yue, M., Lambert, H., Pahon, E., Roche, R., Jemei, S., Hissel, D.: Hydrogen energy systems: a critical review of technologies, applications, trends and challenges. Renew. Sust. Energ. Rev. **146**, 111180 (2021). https://doi.org/10.1016/j.rser.2021.111180
17. Asdrubali, F., Desideri, U.: Chapter 7—high efficiency plants and building integrated renewable energy systems. In: Handbook of Energy Efficiency in Buildings, pp. 441–595. Butterworth-Heinemann, Oxford (2019). https://doi.org/10.1016/B978-0-12-812817-6.00040-1
18. Lu, K., Jiang, X., Yu, J., Tam, V.W., Skitmore, M.: Integration of life cycle assessment and life cycle cost using building information modeling: a critical review. J. Clean. Prod. **285**, 125438 (2021). https://doi.org/10.1016/j.jclepro.2020.125438
19. Di Somma, M., Yan, B., Bianco, N., Graditi, G., Luh, P., Mongibello, L., Naso, V.: Multi-objective design optimization of distributed energy systems through cost and exergy assessments. Appl. Energy **204**, 1299–1316 (2017). https://doi.org/10.1016/j.apenergy.2017.03.105
20. Sheikhi, A., Ranjbar, A.M., Safe, F., Mahmoodi, M.: CHP optimized selection methodology for an energy hub system. In: 2011 10th International Conference on Environment and Electrical Engineering, pp. 1–5 (2011). https://doi.org/10.1109/EEEIC.2011.5874600
21. Zhou, Z., Liu, P., Li, Z., Ni, W.: An engineering approach to the optimal design of distributed energy systems in China. Appl. Therm. Eng. **53**(2), 387–396 (2013). https://doi.org/10.1016/j.applthermaleng.2012.01.067

Chapter 5
Multi-objective Capacity Determination Method of Energy Storage for Smelting Enterprises Considering Wind/Photovoltaic Uncertainty and Clean, Low-Carbon, Economic Indicators

Wenguang Zhu, Wei Wang, Bin Ouyang, Hua Zhang, and Xin Wang

5.1 Introduction

With the increasing consumption of fossil energy and the aggravation of environmental problems, it will be the future trend to gradually replace fossil energy with renewable energy such as wind power and photovoltaic, which is the inevitable way to achieve the "double carbon" goal [1]. Clean energy replacement and industrial process energy saving and carbon reduction are the necessary branches of green and low-carbon transformation for enterprises with high energy consumption and high emission, including electrolytic aluminum enterprises [2]. Firstly, this paper studies the carbon emission from the industry level and establishes the carbon emission measurement method of electrolytic aluminum industry. Then, for the real users in this industry, the total carbon emissions of the users are calculated according to the resource consumption and load data. After that, based on the billing rules of large industrial users, energy storage is configured at large industrial users of electrolytic aluminum that have installed or plan to install new energy generators. The configuration and operation optimization of energy storage are carried out with the goal of maximizing user revenue, the highest proportion of new energy consumption, and the minimum carbon emissions of users.

W. Zhu · W. Wang · H. Zhang · X. Wang
Economic and Technological Research Institute, State Grid Jiangxi Electric Power Co., Ltd, Nanchang City, Jiangxi Province, China

B. Ouyang (✉)
Changsha University of Science and Technology, Changsha City, Hunan Province, China

X. Wang (ed.), *Future Energy*, Green Energy and Technology,
https://doi.org/10.1007/978-3-031-33906-6_5

5.2 Carbon Emission Model of Electrolytic Aluminum Enterprise

The carbon dioxide emission accounting of electrolytic aluminum enterprises includes the carbon emission of fossil fuel combustion in all production systems of the enterprise, the carbon emission of energy used as raw materials, the carbon emission of industrial production process, and the sum of the carbon emission of electric energy and heat consumed by the enterprise [3]. Among them, carbon emissions due to electricity consumption accounted for about 64.8% of the total, followed by thermal energy emissions (including purchased heat and fuel combustion) accounted for about 12.7%. The reaction equation of the essence of electrolytic aluminum process is as follows:

$$2AL_2O_3 + 3C == 4AL + 3CO_2 \tag{5.1}$$

The carbon emission measurement model of electrolytic aluminum enterprises was established by referring to the "Guidelines for Accounting Methods and Reporting of Greenhouse Gas Emissions of electrolytic Aluminum Manufacturers in China," and the calculation formula was as follows:

$$R = R_{\text{burn}} + R_{\text{process}} + R_{\text{elec\&power}} + R_{\text{material}} \tag{5.2}$$

$$\begin{cases} R_{\text{process}} = R_{PFCs} + R_{\text{lim}\,e} = (6500 \times EF_{\text{CF4}} + 9200 \times EF_{\text{C2F6}}) \\ \qquad \times P_{li}/1000 + L \times EF_{\text{lim}\,e} \\ R_{\text{elec\&power}} = AD_{\text{elec}} \times EF_{\text{elec}} + AD_{\text{power}} \times EF_{\text{power}} \\ R_{\text{material}} = EF_{\text{carbon-anode}} \times P_{li} \end{cases} \tag{5.3}$$

In the formula: R_{burn}, R_{process}, R_{elec}, and R_{material} are carbon dioxide emissions from fossil fuel combustion, production processes, electricity, and raw material use, respectively, EF_{CF4} and EF_{C2F6} are carbon emission factor of CF_4 and C_2F_6, L is raw material consumption, $EF_{\text{lim}e}$ is carbon emission factor of lime, AD_{elec}, AD_{power} are purchased electricity and heat, EF_{elec}, EF_{power} are carbon emission factors of electricity and heat consumption, $EF_{\text{carbon - anode}}$ is carbon anode consumption of carbon dioxide emission factor, and P_{li} is aluminum production.

According to the carbon emission accounting standard of the aluminum industry and the current technological level of electrolytic aluminum in the country, the carbon emission accounting is directly converted into 13 tons of carbon dioxide for every ton of aluminum produced. If coal power is replaced by clean energy, 0.96 tons of carbon emissions can be reduced for every 1 MWh of coal power consumption [4]. Therefore, replacing coal power with clean energy is the most effective way to reduce carbon emissions of enterprises with high energy consumption and high emission.

5.3 Multi-objective Collaborative Optimization Model and Constraint Conditions

5.3.1 The Objective Function

Minimum Energy Storage Cost System cost mainly includes capacity cost, power cost, and maintenance cost under the whole life cycle. The system cost can be calculated as follows:

$$\min F_1 = \left(m_p + m_y\right) P_{b.n} + m_B S_n \tag{5.4}$$

In the formula: m_p, m_y, m_B are power, maintenance costs, and storage capacity unit price of energy storge, $P_{b.n}$ is energy storage rated power, and S_n is rated capacity.

Maximum Direct Benefit from Energy Storage The electricity charge paid by industrial users includes two parts: basic electricity charge and electricity rate. The basic electricity bill is multiplied by the maximum monthly amount of the user's demand by the unit price of the basic electricity bill; the electricity rate is equal to the actual electricity consumption times the unit price.

$$\max F_2 = C_{f.j} + C_{x.j} \tag{5.5}$$

$$\begin{cases} C_{f.j} = \sum\limits_{j=1}^{M} m\left(P_{L\max.j} - P_{\max.j}\right) \\ C_{x.j} = -\sum\limits_{j=1}^{M}\sum\limits_{k=1}^{N}\sum\limits_{t=1}^{96}\left[e(t)\left(P_{bc.t} + P_{bd.t}\right)h\right] \end{cases} \tag{5.6}$$

In the formula: $C_{f.j}$, $C_{x.j}$ are the defensive revenue and peak cutting and grain filling revenue of the JTH month, respectively; m is basic electricity price; M is number of months in the simulation period; $P_{L\max.j}$, $P_{\max.j}$ are maximum user load and monthly demand defense value; N is the number of days in month M; and $e(t)$ is time-of-use electricity price.

Maximum Carbon Emission According to the latest Carbon Emission Trading Management Measures issued in China, industrial users have a certain amount of free carbon dioxide quota, and the excess part needs to be introduced into the paid quota according to the requirements. Therefore, industrial users can consume more new energy through energy storage to reduce carbon emissions from electricity consumption. Carbon emission reduction is quantified as economic quantity, as shown in Eq. (5.7).

$$\max F_3 = F_{\text{fine}} \times \Delta Q_{\text{ex}} \tag{5.7}$$

In the formula: ΔQ_{ex} is carbon emission reductions.

5.3.2 *Constraints*

Power Equality Constraint [5, 6]

$$\begin{cases} P_{L.t} + P_{b.t} = P_{pv.t} + P_{w.t} + P_{\text{net}} \\ -P_{b.n} \le P_{b.t} \le P_{b.n} \\ 0.17 \le SOC_{b.t} \le 0.95 \\ Q_{i.t} \le Q_{\max} \\ P_{b.n} = S_n/\eta_1 \\ 0 \le P_{L.t} + P_{B.t} - P_{pv.t} - P_{w.t} \le P_{L\max} \end{cases} \tag{5.8}$$

In the formula: $P_{L.\,t}$ is power of load, $P_{b.\,t}$ is energy storage power, $P_{pv.\,t}$ is photovoltaic power, $P_{w.\,t}$ is wind power, P_{net} is grid power, $SOC_{b.\,t}$ is the charged state, $Q_{\max}$ is maximum switching power, η_1 is performance factors, and $P_{L\max}$ is maximum net load power.

5.4 Model Solving Algorithm

5.4.1 *Energy Storage Configuration Optimization Model*

In order to comprehensively consider the influence of the randomness and volatility of the new energy of wind and light on the energy storage capacity configuration in the wind-light-storage system, an energy storage configuration optimization model under the user-side new energy access is constructed. Based on the real data of local historical wind and optical resources as well as the fixed rated power of wind and optical generating units, the wind speed and solar illumination intensity of 1000 days with an interval of 15 min were randomly generated by Monte Carlo, and then 1000 output curves were calculated by plugging them into the output model of wind and optical generating units. The constraint conditions are Eq. (5.8). The energy storage capacity and power of 1000 optimization cycles are optimized and solved, and their mean values are taken as the rated capacity and power of energy storage final optimization.

$$\max F = 1/2 \times (\alpha_1 F_1 - \alpha_2 F_2) + 1/2 \times F_3 \tag{5.9}$$

5.4.2 Energy Storage Operation Optimization

Firstly, the rated power of new wind and light energy connected by the user is determined, and the output of new energy is simulated randomly through Weibull distribution and beta distribution. Considering the influence of drifting clouds on photovoltaic, random fluctuations are generated on the original basis to simulate the scenario where clouds cover the sunlight. According to the impact characteristics of electric load of electrolytic aluminum [7], the impact load is introduced as the forecast load at certain moments of typical daily load to optimize the real-time power of energy storage operation in the day, so as to obtain the maximum energy storage income and carbon emission reduction income.

$$\max F = F_2 + F_3 \tag{5.10}$$

5.4.3 Energy Storage Optimization Process

The constant volume optimization process of energy storage is as follows:

Step 1: Establish the energy storage configuration optimization model, and set the coefficients of the objective function.

Step 2: Determine the rated power of the new energy that has been or will be installed, take day as the optimization period, and randomly generate the output curve for 1000 days based on their respective characteristics. Then, the user's historical load data is analyzed to determine a typical daily load for optimization. The constraint conditions are Eqs. (5.8)–(5.10). The optimal rated capacity and power of energy storage under the new energy rated power are obtained by solving the energy storage configuration optimization model.

Step 3: Taking the capacity and power obtained above as constraints, an optimization model of energy storage operation within a day is established. Based on the determined rated power of new energy, the theoretical output power is randomly generated according to different distribution characteristics. Combined with the typical daily load of the user, the intraday operation model of energy storage is solved to obtain the real-time power of energy storage with the goal of maximizing the energy storage revenue and carbon emission reduction revenue.

Step 4: Determine whether the rate of new energy consumption is greater than the threshold. If the rate of new energy is less than the threshold, α_1 will be reduced and α_2 will be increased. Next, continue to judge whether the new energy consumption rate is greater than the threshold. If it's still less than, then repeat step 2 to step 4 until the solution is successful to obtain the energy storage capacity and power.

Table 5.1 Cost parameters of energy storage batteries

Energy storage type	Storage costs		
	Power	Capacity	Maintenance
Lithium iron phosphate battery	60,000	160,000	15,500

5.5 Example Simulation and Result Analysis

5.5.1 *The Example System*

For example, the power supply voltage level of the electrolytic aluminum enterprise is 10 kV. Under this voltage level, the local basic electricity price is 39 yuan/kW, and the pricing rule of two-part system for local large industry is adopted. The price of electricity is the real TOU price in a province of China. The energy storage device adopts the most common lithium iron phosphate battery in the market, and its related parameters are shown in Table 5.1.

5.5.2 *Energy Storage Optimization*

Before clean energy replacement, an electrolytic aluminum enterprise in China produced 110 tons of electrolytic aluminum on average every day, and the daily carbon dioxide emission of the user was calculated to be 1474.7 tons, among which the electricity consumption was 989.25 MWh. Then, 949.68 tons of carbon dioxide was produced due to the consumption of electric energy, accounting for 64.4% of the total carbon dioxide emission of the enterprise. It is assumed that the rated power of wind power installed by the user is 25 MW and that of photovoltaic power is 30 MW. The output characteristics of new energy of wind and light are fully taken into account. On this basis, energy storage is configured for the user. The following is the optimization result under 1000 optimization cycles. The final result is obtained by k-means clustering after eliminating abnormal data, as shown in Fig. 5.1 and Table 5.2 below.

Partial optimization results and corresponding user carbon emission data under typical daily load are shown in Table 5.3.

The real-time optimization results of energy storage are shown in Fig. 5.2. The energy storage device charges when the output of new energy is large and discharges at the peak of electricity price to increase user income. The daily SOC of the battery storage device is maintained between 0.2 and 0.95 in real time, and the initial and final values are equal, and the power exchange is twice the rated capacity. The optimization strategy in this paper keeps the energy storage battery in a very healthy running state and prolongs the service life of the energy storage device. In this case, the output power of new energy is 310.798 MWh, and the consumption rate is

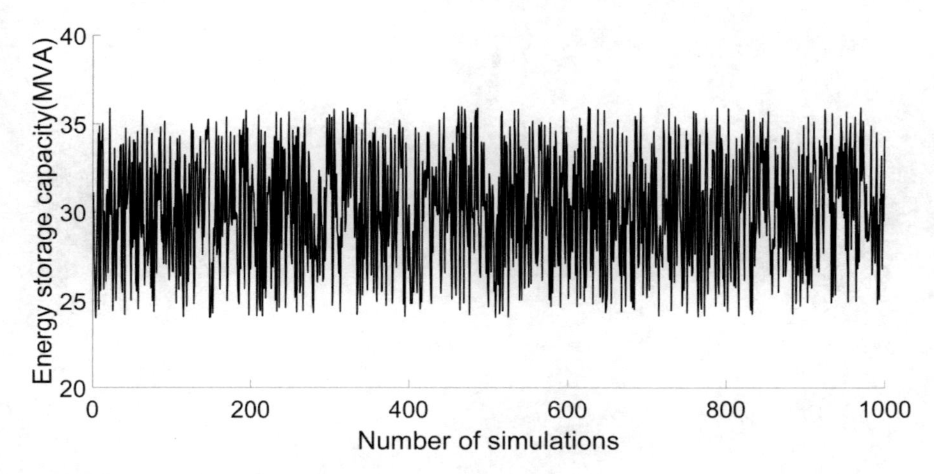

Fig. 5.1 Multiple optimization results

Table 5.2 Optimization results of energy storage configuration

Energy storage parameters		Cost (yuan)	Benefits (yuan per month)	Cost recovery period
Capacity (MVA)	Power (MW)			
23.946	11.973	47,353,000	519,220	7.6

Table 5.3 Partial optimization results

Cases	Wind power permeability	Photovoltaic permeability	Energy storage power	Energy storage capacity	Payback period	Carbon emissions (ton)	Carbon emission reduction (ton)
1	0.181	0.157	10.14	20.28	7.89	627.9	321.7
2	0.213	0.158	10.93	20.86	6.97	597.9	351.7
3	0.202	0.159	14.36	28.72	7.28	606.6	343.0
4	0.214	0.161	10.64	21.28	6.60	593.7	355.9
5	0.170	0.163	13.87	27.74	7.92	633.9	315.7
6	0.159	0.163	9.43	18.86	7.66	644.1	305.5
7	0.229	0.163	10.14	20.28	7.87	577.5	372.1
8	0.237	0.162	10.93	21.86	7.72	570.4	379.2
9	0.228	0.161	10.96	21.92	8.43	580.5	369.1
10	0.190	0.159	10.02	20.04	6.73	618.0	331.6

100%, which reduces the emission of carbon dioxide by 298.37 tons, accounting for 20.87% of the total emission.

For industrial users, the long-term goal of installing energy storage is to consume a high proportion of new energy to reduce their total carbon emissions, followed by the direct economic benefits brought by energy storage. Increase the rated power of wind power and photovoltaic to 30 MW and 35 MW, and increase the permeability

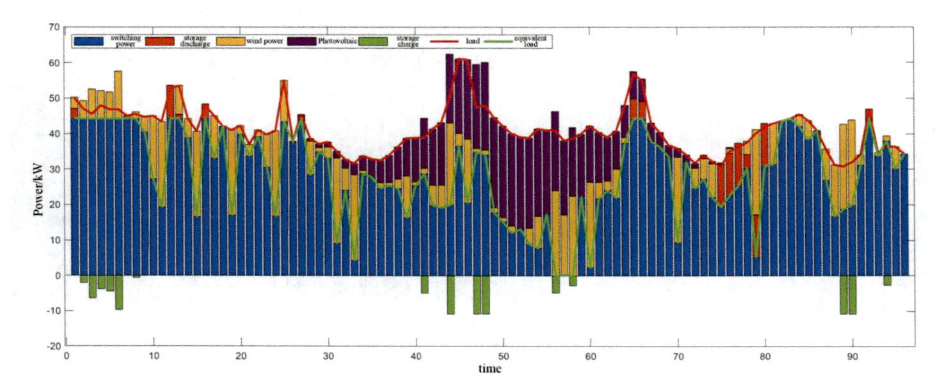

Fig. 5.2 Timing optimization result

to 24.57% and 21.03%, respectively. At this time, there are many times when the total power generation of new energy is greater than the load. If there is no energy storage device, the surplus can only be wasted. Through the above constant volume optimization of energy storage, it is concluded that the rated power of energy storage should be 20 MW and the capacity should be 40 MWh under this permeability.

After the proportion of new energy is increased, if it is absorbed by the system itself, the available new energy is 451.14 MWh, and the abandoned power is 41.51 MWh. The total carbon emission of the power load of the customer is reduced from the original 949.68 tons to 556.44 tons. After the energy storage optimization method in this paper is adopted for optimization, the intraday operation optimization result is shown in Fig. 5.2. The new energy consumption rate is 100%, and the carbon dioxide emission of power load is reduced to 516.59 tons, which is 4.2% of the total carbon emission on the basis of not adding energy storage, saving the carbon emission cost of 3119.4 Yuan. The results show that if the penetration rate of new energy continues to increase, the energy storage optimization strategy in this paper can effectively improve the consumption of new energy and reduce carbon emission.

5.6 Conclusion

The energy storage constant capacity optimization strategy proposed in this paper can fully consider the uncertainty of new energy sources and the potential carbon emission market to optimize the rated power and capacity of energy storage at the industrial user side.

Acknowledgments This work was financially supported by the Science and Technology Project of State Grid Jiangxi Electric Power Co., Ltd. Item number: 521825220001.

References

1. Zhang, S.X., Wang, D.Y., Cheng, H.Z.: Key technologies and challenges of low-carbon integrated energy system planning for carbon emission peak and carbon neutrality. Autom. Electr. Power Syst. **46**(08), 189–207 (2022)
2. Li, B., Chen, M.Y., Zhong, H.W.: A review of long-term planning of new power systems with large share of renewable energy. In: Proceedings of the CSEE, pp. 1–27 (2022)
3. Wang, W., Wu, J.J., Ge, Y.P.: Carbon Emission Accounting Method and Strategy Analysis under the Background of Double Carbon. Taking Copper and Aluminum Industry as an Example.Nonferrous Metals (Extractive Metallurgy) (04), pp. 1–11 (2022)
4. He, Y., Xing, Y.T., Ji, Y.J.: On influential factors and regional difference in carbon emissions from power industry at home in China. J. Saf. Environ. **20**(06), 2343–2350 (2020)
5. Zhao, Y.T., Wang, H.F., He, B.T.: Optimization strategy of configuration and operation for user-side battery energy storage. Autom. Electr. Power Syst. **44**(06), 121–128 (2019)
6. Chen, H.H., Du, H.H., Zhang, R.F.: Optimal capacity configuration and operation strategy of hybrid energy storage considering uncertainty of wind power. Electr. Power Autom. Equip. **38**(08), 174–182 (2018)
7. Guo, C., Zhu, R.L., Meng, X.: Modeling method of electrolytic aluminum impact load based on gray wolf optimization algorithm. Electr. Eng. **11**, 65–70 (2021)

Chapter 6
The Optimal Allocation Strategy of Pumped Storage for Boosting Wind/Solar Local Consumption

Wenru Liang, Linwei Sang, Xiaolin Luo, Xin Sui, Yong Luo, Hengyu Gan, Li Huang, and Yinliang Xu

6.1 Introduction

6.1.1 A Subsection Sample

The utilization of renewable energy is a promising solution to ensure future energy security and cope with the problem of global climate change. Although the development of renewable energy sources is changing rapidly, which is of great benefit to the realization of environmental protection, the proportion of renewable energy sources in power generation is not high. Wind and photovoltaics have the characteristics of intermittency, fluctuation, and randomness, which makes it difficult to consume [1]. The microgrid can fully promote the large-scale access of distributed generators and renewable energy sources, which realizes the reliable supply of load [2]. Due to the fragmentation of various energy sources, comprehensive and cascade utilization is required. Meanwhile, a large amount of wind, photovoltaic and hydro-energy curtailment exists owing to the serious lack of flexible resources. Therefore, it is necessary to make full use of the abundant domestic hydro-energy, wind, and photovoltaic and optimize the allocation of clean energy to build a clean, low-carbon energy system.

Wind energy and solar energy have the advantages of wide distribution, large reserves, and short infrastructure cycle. However, it is unstable in a short time

W. Liang · L. Sang · Y. Xu
Tsinghua University, Shenzhen, China
e-mail: liangwr20@mails.tsinghua.edu.cn; xu.yinliang@sz.tsinghua.edu.cn

X. Luo · X. Sui (✉)
China Institute of Water Resources and Hydropower Research, Beijing, China

Y. Luo · H. Gan · L. Huang
Guizhou Wujiang Hydropower Development Ltd., Guiyang, China

X. Wang (ed.), *Future Energy*, Green Energy and Technology,
https://doi.org/10.1007/978-3-031-33906-6_6

and its output is difficult to store in a large-scale power system. Hydro-energy has superiority in abundant installed capacity, high start-stop speed, flexible operation, and load matching. The complementary development of wind-solar-pumped-storage hybrid-energy system has gradually become an important way to deal with the randomness, intermittency, and fluctuation of renewable energy sources. The fluctuation of wind and photovoltaic can be stabilized by the adjustment of hydropower, which provides support for the consumption of renewable energy sources in the power grid. Meanwhile, the seasonal complementary characteristics of wind and photovoltaic to hydropower can provide power support during dry seasons and peak shaving capacity during wet seasons for hydropower. Therefore, the hybrid-energy complementary development of wind-solar-pumped-storage system and its capacity allocation are of great significance to solving the problem of wind-photovoltaic consumption.

In this regard, scholars have carried out a great number of studies. At present, domestic and abroad researches on the allocation of wind-photovoltaic-hydropower complementary capacity mostly focus on the reliability of the power system and the total cost of system construction and operation. In [3], it aims to increase the stability of the system and comprehensively considers the constraints such as utilization for wind-photovoltaic resource, the system rated capacity, and the discharge depth. A wind-photovoltaic capacity allocation optimization model for hybrid system planning is proposed. In [4], the minimum probability of insufficient power supply of the wind-solar system is taken as the optimization objective considering the power supply reliability. The capacity allocation of the wind-solar system is decided based on the minimum unit cost of the power system. However, the mentioned models are more practical in a small-scale power grid, which is difficult to obtain an optimal solution in the complementary hybrid-energy system. In a complex system, the complementarity of wind and photovoltaic and the output characteristics of hydropower should be fully considered to suppress the fluctuation of renewable energy sources. In [5], considering the fluctuation of renewable energy sources, the capacity of the wind-solar-storage hybrid power system is optimally configured based on prediction error compensation and fluctuation suppression. The seasonal characteristics of wind and solar are also taken into account to define the output fluctuation index and to determine the capacity allocation in [6, 7]. However, the deviation between wind-photovoltaic output and the load is ignored in the references, which cannot effectively track the load. In [8], a dispatch strategy for the photovoltaic-wind-pumped-storage hybrid power system is proposed to reduce the power curtailment of wind and photovoltaic. In [9], the complementarity index to measure the complementarity of the wind-solar-hydro output of the hybrid system is defined. With the goal of maximizing the index, the optimal site and capacity of a wind farm and photovoltaic station supported by a hydropower station is decided. In addition to determining the relevant indicators, a corresponding algorithm must be used to determine the optimal ratio of wind and solar installed capacity. In the past, different methods have been used based on different objectives: particle swarm optimization (PSO) [10], genetic algorithm [11], mixed integer nonlinear programming [12], and mixed simulated annealing tabu search algorithm [13].

These methods above offer different allocation options and the final result is decided by the decision-maker.

In summary, the wind-solar resources have the reverse distribution with the power load and the hydropower has great potential to be developed and utilized. Therefore, it is necessary to study the allocation strategy of wind-solar-pumped-storage resources that considers the local consumption of renewable energy sources. In this paper, pumped storage is taken as an example. First, based on the actual wind-solar output and load data of a certain area in Sichuan, a cluster analysis is carried out to obtain a typical scene of the area for 1 year. Furthermore, a wind-solar-pumped-storage energy ratio planning strategy is proposed considering the local consumption. The influence of different photovoltaic ratios and uncertainties on the optimization results is analyzed.

6.2 Model for Power Output

6.2.1 Model for Photovoltaic

The power fluctuation model of photovoltaic is

$$P_{pv} = P_{pv1}\eta_{inv}\eta_{loss}\eta_{ref}\left[1 + k_T\left(T_0 - T_{ref}\right)\right]\frac{G_\alpha}{1000} \tag{6.1}$$

where P_{PV1} is rated power of the photovoltaic cell, η_{inv} is the efficiency of inverter, η_{loss} is the loss efficiency of the photovoltaic cell, η_{ref} is the efficiency of photovoltaic cell at the reference temperature, k_T is the power-temperature coefficient of the PV panel, T_0 is the operating temperature of photovoltaic, T_{ref} is the reference temperature of photovoltaic, and G_α is the hourly average value of the solar radiation on the surface.

6.2.2 Model for Wind Turbine

The output of wind turbine mainly depends on the wind speed. The relationship between the power output P_w and the wind speed v can be approximately described by a piecewise function, which is

$$P_W = \begin{cases} 0 & ,v < v < v_{in}\text{或}v \geq v_{out} \\ P_{Wr}\dfrac{v - v_{in}}{v_r - v_{in}} & ,v_{in} \leq v < v_r \\ P_{Wr} & ,v_r \leq v < v_{out} \end{cases} \tag{6.2}$$

where v_r and P_{Wr} indicate the rated wind speed and rated power of the wind turbine. v_{in} and v_{out} are the cut-in and cut-out wind speed, respectively.

6.2.3 Model for Wind Turbine

The operation model of the pumped-storage power station is:

$$\begin{cases} 0 \leq P_{PSH}(t) \leq \gamma_H P_{PS\max} \\ 0 \leq P_{PSP}(t) \leq \gamma_P P_{PS\max} \\ \gamma_H + \gamma_P \leq 1 \end{cases} \tag{6.3}$$

where $P_{PS\max}$ is the maximum installed capacity of the reversible pump-turbine, $E_{PS\max}$ is the power generation corresponding to the maximum volume of the upstream storage capacity, and $P_{PS}(t)$ is the actual power at time t. The pumped-storage power station is releasing water to generate electricity when $P_{PS}(t)$ is greater than 0.

The constraint of upstream storage capacity is:

$$c_{PS\min} E_{PS\max} \leq E_{PS}(t) \leq c_{PS\max} E_{PS\max} \tag{6.4}$$

where $E_{PS}(t)$ is the upstream storage capacity and $c_{PS\min}$ and $c_{PS\max}$ are minimum and maximum storage capacity ratios, respectively.

The constraint on the relationship between the storage capacity and the actual operating power is:

$$E_{PS}(t) = E_{PS}(t-1) + (\eta_P P_{PSP}(t) - \eta_H P_{PSH}(t))\,\Delta T \tag{6.5}$$

The constraint on the water inflow and yield is:

$$\sum_{t=1}^{N_t} (P_{PSH}(t) - P_{PSP}(t)) = 0 \; t \in \{1, \ldots, N_t\} \tag{6.6}$$

6.3 Model for Boosting Wind/Solar Local Consumption

The total cost of the pumped-storage power station is mainly composed of the installed capacity cost, the storage capacity construction cost, and the regular maintenance cost, which is

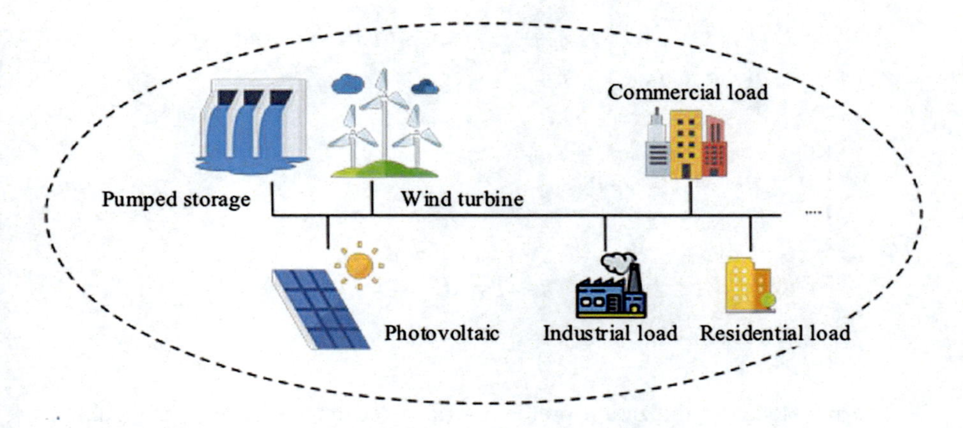

Fig. 6.1 Wind-photovoltaic local consumption

$$NPC_{PS} = C_{PScap}V_{PScap} + C_{PSpower}P_{PSpower} + \frac{C_{reppc}P_{PSpower}}{(1+r)^{T_{PS}}} + \sum_{n=1}^{T_a} \frac{C_{OMpv}V_{PScap}+C_{OMpp}P_{PSpower}}{(1+r)^n} \tag{6.7}$$

where V_{PS_cap} is the volume of the upstream storage capacity, P_{PS_power} is the installed capacity of the reversible pump-turbine, C_{PS_cap} is the price per cubic meter of the upstream storage capacity, C_{PS_power} is the price per kilowatt of installed capacity of the turbine, C_{rep_pc} is the replacement cost of the turbine, T_{PS} is the life cycle of the turbine, C_{OM_pv} is the operation and maintenance cost per cubic meter of the upstream storage capacity, and C_{OM_pp} is the operation and maintenance cost per kilowatt.

The planning model based on load matching is constructed, considering that local consumption of renewable energy could be realized through the joint planning of wind, solar, water, and thermal. The schematic diagram is shown in Fig. 6.1.

6.3.1 Deterministic Optimization

The planning model of local consumption for wind, solar, and water mainly utilizes the regulating ability of pumped storage to promote wind and solar to match the intraday characteristics of loads. The deterministic optimization model is carried out ignoring the prediction error of wind and solar. The objective function mainly includes two parts, planning cost and operation cost.

The deterministic optimization model is as follows:

$$\begin{aligned}
&\min NPS_{PS}(P_{PS\max}, E_{PS\max}) + \sum_{t} c_{PSH} P_{PSH} + c_{PSP} P_{PSP} \\
&s.t. P_g + {P_{sw}}^{ac} + P_{PSH} + P_{PSP} = P_{\text{load}} \\
&\quad \sum {P_{sw}}^{ac} \le (1 - \text{drop}) \sum P_{sw} \\
&\quad \begin{cases} 0 \le P_{PSH}(t) \le \gamma_H P_{PS\max} \\ 0 \le P_{PSP}(t) \le \gamma_P P_{PS\max} \\ \qquad \gamma_H + \gamma_P \le 1 \end{cases} \\
&\quad c_{PS\min} E_{PS\max} \le E_{PS}(t) \le c_{PS\max} E_{PS\max} \\
&\quad E_{PS}(t) = E_{PS}(t-1) + (\eta_P P_{PSP}(t) - \eta_H P_{PSH}(t)) \Delta T \\
&\quad \sum_{t=1}^{N_t} (P_{PSH}(t) - P_{PSP}(t)) = 0 \quad t \in \{1, \ldots, N_t\}.
\end{aligned} \tag{6.8}$$

The mentioned optimization problem is a mixed integer linear programming model, which could be directly solved by the commercial solver GUROBI.

6.3.2 Stochastic Optimization

The installed capacity allocation of the pumped-storage power station in view of the uncertainty is obtained on the basis of deterministic optimization, in which the stochastic scenario is considered where there is a deviation between the actual wind and solar output and the predicted value. Stochastic optimization is utilized to configure for the pumped-storage power station.

The stochastic optimization model is as follows:

$$\begin{aligned}
&\min NPS_{PS}(P_{PS\max}, E_{PS\max}) + \sum_{s \in S} \sum_{t} (c_{PSH} {P_{PSH}}^s + c_{PSP} {P_{PSP}}^s) \\
&s.t. {P_g}^s + {P_{sw}}^{ac,s} + P^s{}_{PSH} + {P_{PSP}}^s = P_{\text{load}} \\
&\quad \sum {P_{sw}}^{ac,s} \le (1 - \text{drop}) \sum {P_{sw}}^s \\
&\quad \begin{cases} 0 \le {P_{PSH}}^s(t) \le {\gamma_H}^s P_{PS\max} \\ 0 \le {P_{PSP}}^s(t) \le {\gamma_P}^s P_{PS\max} \\ \qquad {\gamma_H}^s + {\gamma_P}^s \le 1 \end{cases} \\
&\quad c_{PS\min} E_{PS\max} \le {E_{PS}}^s(t) \le c_{PS\max} E_{PS\max} \\
&\quad {E_{PS}}^s(t) = {E_{PS}}^s(t-1) + (\eta_P {P_{PSP}}^s(t) - \eta_H {P_{PSH}}^s(t)) \Delta T \\
&\quad \sum_{t=1}^{N_t} ({P_{PSH}}^s(t) - {P_{PSP}}^s(t)) = 0 \quad t \in \{1, \ldots, N_t\}.
\end{aligned} \tag{6.9}$$

The model could be directly solved via the commercial solver GUROBI.

6.3.3 Robust Optimization

Furthermore, consider the worst scenario where the deviation between the actual wind and solar output and the predicted value occurs. Robust optimization is utilized to configure the pumped-storage power station to obtain a conservative capacity allocation scheme considering uncertainty.

A max-min robust optimization model is carried out, in which the inner layer is the planning model and the outer layer is the optimization model for uncertainty variables. The robust optimization model is obtained as shown:

$$\begin{aligned}
&\max_{\tilde{P}_{sw},\tilde{P}_{sw}\in U} \min \quad c_{ps}C_{ps} + c_h C_h \\
&s.t. E_{ps}(t) = E_{ps}(t-1) - P_{\text{dis}}(t) + P_{ch}(t) \\
&\quad t = 1,\ldots,T\ (\mu_{ps}) \\
&\quad c_{ps,\min}C_{ps} \le E_{ps} \le c_{ps,\max}C_{ps} \quad (\lambda_{ps,-},\lambda_{ps,+}) \\
&\quad E(0) = c_0 C_{ps} \\
&\quad E(t) = E(t+23) \quad t = 1, 25, ..24^*8+1 \\
&\quad \mu_{\text{day}} 0 \le P_{\text{dis}} \le c_{\text{dis}}C_{ps} \quad (\lambda_{\text{dis}}) \\
&\quad 0 \le P_{ch} \le c_{ch}C_{ps} \quad (\lambda_{ch}) \\
&\quad c_{h,\min}C_h P_{\text{hydro}} \le P_h \le c_{h,\max}C_h P_{\text{hydro}} \quad (\lambda_{h,-},\lambda_{h,+}) \\
&\quad 0 \le C_h, C_{ps}\tilde{P}_{sw} + P_{\text{dis}} - P_{ch} + P_h = \tilde{P}_d.
\end{aligned} \tag{6.10}$$

The inner min problem is converted into a max problem by the dual theory to solve the worst scenario, which is shown as

$$\begin{aligned}
&\max \quad \mu_{\text{bal}}^T (P_{sw} - P_d) \\
&s.t. \mu_{ps}(t) - \mu_{ps}(t+1) + \lambda_{ps,-}(t) - \lambda_{ps,+}(t) = 0, t = 1,\ldots,T-1 (E) \\
&\quad \mu_{ps}(T) + \lambda_{ps,-}(T) - \lambda_{ps,+}(T) = 0 \\
&\quad \mu_{ps}(t) - \lambda_{\text{dis}}(t) + \mu_{\text{bal}}(t) \le 0, t = 1,\ldots,T\ (P_{\text{dis}}) \\
&\quad -\mu_{ps}(t) - \lambda_{ch}(t) - \mu_{\text{bal}}(t) \le 0, t = 1,\ldots,T\ (P_{ch}) \\
&\quad \Big[-c_0\mu_{ps}(1) - c_{ps,\min}\lambda_{ps,-} + c_{ps,\max}\lambda_{ps,+} \\
&\quad + c_{\text{dis}}\lambda_{\text{dis}} + c_{ch}\lambda_{ch} \Big] \le \text{cost}_{ps}\,(C_{ps}) \\
&\quad -c_{h,\min}\lambda_{h,-}P_{\text{hydro}} + c_{h,\max}\lambda_{h,+}P_{\text{hydro}} \le \text{cost}_h\,(C_h) \\
&\quad \lambda_{h,-} - \lambda_{h,+} + \mu_{\text{bal}} \le 0\,(P_h).
\end{aligned} \tag{6.11}$$

The mentioned optimization problem is a linear programming model, which can be directly solved by the commercial solver GUROBI.

6.4 Case Studies

Considering the different access of wind turbines and photovoltaics in different regions, the proportion of photovoltaics is defined to describe the proportional relationship between the installed capacity of wind turbines and photovoltaics. The full name of photovoltaic ratio portion is the ratio of photovoltaic to wind and solar power, which refers to the ratio of the installed capacity of photovoltaic power plants to the total installed capacity of wind turbines and photovoltaics. The value is also between 0 and 1. The specific calculation method is as follows:

$$\text{portion} = \frac{C_{\text{solar}}}{C_{\text{solar}} + C_{\text{wind}}} \tag{6.12}$$

where C_{solar} is the installed capacity of photovoltaics and C_{wind} is the installed capacity of wind turbines.

In the planning application, this paper focuses on the wind and solar access capacity supported by the installed capacity of the unit pumped-storage power station in different scenarios. Therefore, the ratio of pumped-storage and wind-photovoltaic energy is defined, which represents the ratio of the installed capacity of pumped storage to the installed capacity of wind and solar it supports. Specifically, it is shown in as follows:

$$\text{ratio} = \frac{P_{PS\max}}{C_{\text{solar}} + C_{\text{wind}}} \tag{6.13}$$

The fluctuation rate of wind-photovoltaic-hydropower Var_{re} is defined. The value is between 0 and 1. The specific calculation method is as follows:

$$\begin{aligned} \text{Var}_{re} &= \frac{|P_{re} - P_{\text{avg}}|}{P_{\text{avg}} T} \le \delta_{\max} \\ P_{\text{avg}} &= \sum P_{re}/T \end{aligned} \tag{6.14}$$

where P_{re} is the bundled delivery power of renewable energy, P_{avg} is the mean valve of power output, and $\delta_{\max}$ is the maximum output fluctuation rate of wind-PV-pumped-storage hybrid-energy system.

The proposed algorithm is implemented using the output of photovoltaic and wind turbine and load data of Sichuan Province in the planning simulation platform of China Electric Power Research Institute. Typical daily type is clustered based on KMEANS. On the basis of cluster analysis, the allocation planning scheme and the installed capacity ratio of pumped-storage energy to wind-photovoltaic with local consumption are considered. Parameter assumptions for the proposed planning model are shown in Table 6.1.

Table 6.1 Parameter assumption

Parameter	Value
C_{PS_cap} (m.u.)	1.4
C_{PS_power} (m.u.)	2.1
C_{rep_pc} (m.u.)	0.014
C_{OM_pv} (m.u.)	0.021
$c_{PS\min}$	0.2
$c_{PS\max}$	0.95

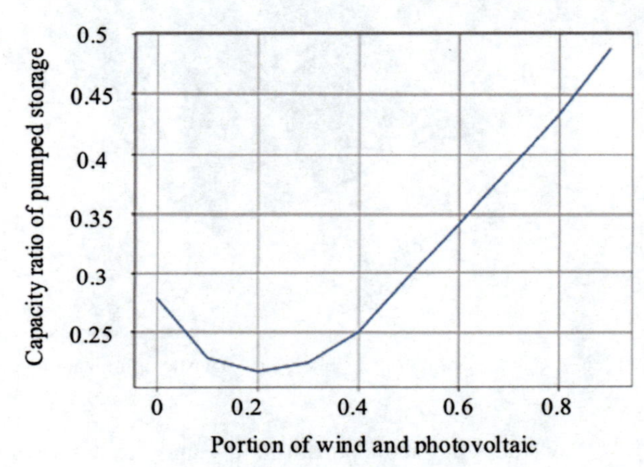

Fig. 6.2 Capacity of pumped storage in deterministic optimization

6.4.1 Deterministic Optimization

Deterministic optimization is carried out on the basis of ignoring prediction errors of wind and solar. The ratio of thermal power access is set to be 30%, and the ratio of the maximum abandonment of wind and solar power is set to be 10%. On this basis, the installed capacity allocation of pumped storage with different ratios for wind and solar is studied in this paper, which is shown in Fig. 6.2.

In deterministic optimization, when the portion for wind and solar is 0.2, the minimum ratio of installed capacity for pumped hydro-energy and wind turbine and photovoltaic occurs, which is 1:3.20.

6.4.2 Stochastic Optimization

The uncertainty of wind and solar refers to the fluctuation range of the prediction deviation of scenery, which is set to be 0–20%. The expected value is minimized by sampling typical scenarios. Other parameters are consistent with the deterministic optimization. The ratio of thermal power access is set to be 30%, and the ratio of the maximum abandonment of wind and solar power is set to be 10%. The planning result is shown in Fig. 6.3.

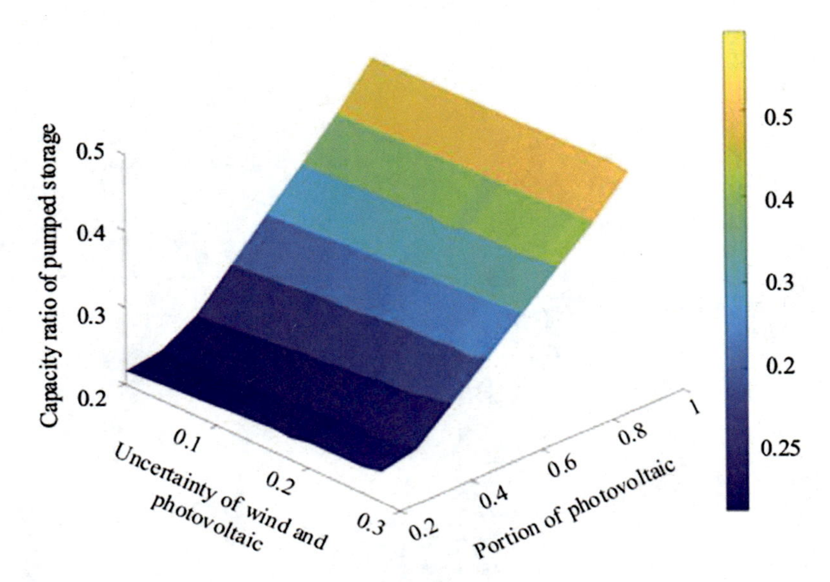

Fig. 6.3 Capacity of pumped storage in stochastic optimization

When the wind-solar portion is 0.4 and the wind-solar uncertainty is 10%, the maximum ratio of the installed capacity for pumped storage and wind-solar capacity is 1:2.65. When the wind-solar portion is 0.4, and the wind-wind uncertainty is 15%, the ratio of the installed capacity for pumped storage and wind-solar capacity is 1:2.61. With the increase of wind-solar uncertainty, the installed capacity of pumped hydro storage increases accordingly.

6.4.3 Robust Optimization

The uncertainty of wind and solar is set to 0–20%. Robust optimization considers the installed capacity allocation in the worst scenario through min-max modeling. Therefore, the results are more conservative. The other parameters are also consistent with the deterministic optimization. The ratio of thermal power access is set to 30%, and the ratio of the maximum abandonment of wind and solar power is set to 10%. The optimization results of robust optimization are shown in Fig. 6.4.

When the wind-solar portion is 0.4 and the wind-solar uncertainty is 10%, the maximum ratio of the installed capacity for pumped storage and wind-solar capacity is 1:2.50. When the wind-solar portion is 0.4, and the wind-wind uncertainty is 15%, the ratio of the installed capacity for pumped storage and wind-solar capacity is 1:2.37. Similarly, with the increase of wind-solar uncertainty, the installed capacity of pumped hydro storage increases accordingly.

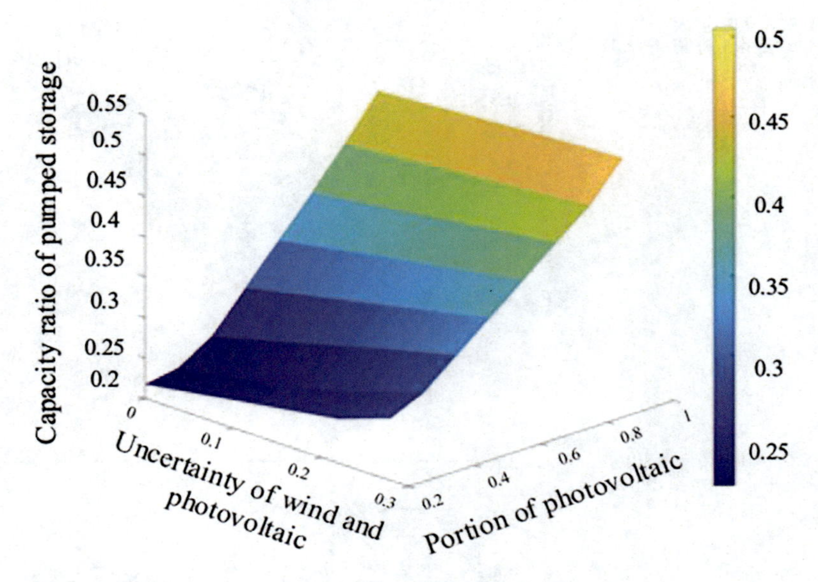

Fig. 6.4 Capacity of pumped storage in robust optimization

Table 6.2 Comparison of proposed methods

(Uncertainty, Portion)	Deterministic optimization	Stochastic optimization	Robust optimization
(0.05, 0.4)	1:2.79	1:2.72	1:2.62
(0.05, 0.6)	1:2.04	1:1.98	1:1.94
(0.1, 0.4)	1:2.79	1:2.65	1:2.50
(0.1, 0.6)	1:2.04	1:1.94	1:1.90
(0.15, 0.4)	1:2.79	1:2.61	1:2.37
(0.15, 0.6)	1:2.04	1:1.91	1:1.86

6.4.4 Comparative Analysis

Furthermore, the comparison of stochastic optimization and robust optimization is carried out, which is shown in Table 6.2. The capacity allocation of robust optimization is significantly higher than that of stochastic optimization, which is more conservative.

The optimal allocation results of the two methods with different uncertainties and ratios are shown in Fig. 6.5. The ratio results of the installed capacity for deterministic optimization, stochastic optimization, and robust optimization are compared. In the case of ignoring the uncertainty of wind and solar, the most optimistic ratio of the installed capacity for pumped storage and wind and solar capacity is obtained and the unit of pumped storage can absorb the most wind-solar capacity.

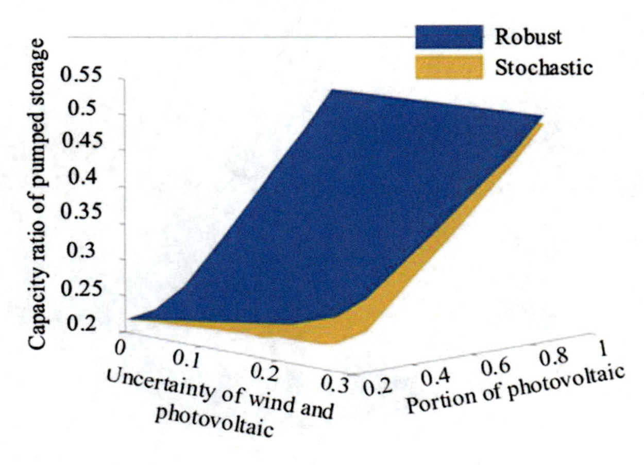

Fig. 6.5 Comparison of stochastic optimization and robust optimization

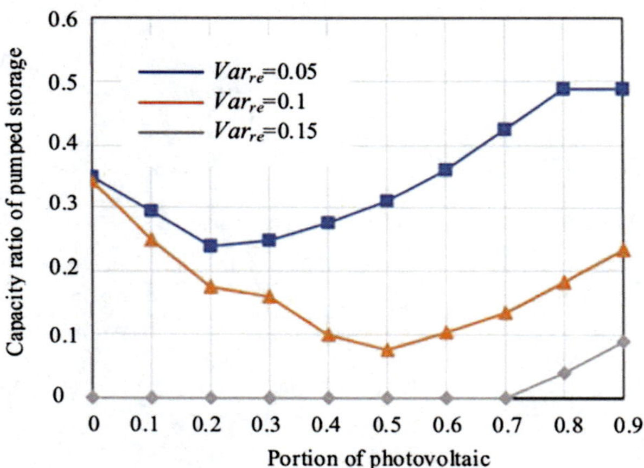

Fig. 6.6 Allocation results for boosting bundled delivery

Meanwhile, the capacity of robust optimization is the worst with the least wind-solar capacity consumption of unit of pumped storage.

6.4.5 Bundled Delivery Boosting of Wind/Solar

Furthermore, in order to cope with the intermittency and uncertainty of wind and photovoltaic, the power supply and energy storage characteristics of pumped-storage station proposed in this paper could also be implemented for boosting wind/solar stable transmission and realizing the complementary development the multi-energy system. The ratio of pumped-storage allocation per unit of wind-photovoltaic with different portion of wind-photovoltaic is shown in Fig. 6.6.

The allocation capacity of pumped storage decreases and then rises with the increase of the proportion of photovoltaics, which illustrates that the complementarity of wind and solar increases and then decreases. In addition, the greater the allowable fluctuation rate of wind and solar output, the lower the pumped-storage allocated capacity will be.

6.5 Conclusion

The optimal allocation strategy of the system in which local consumption of wind and photovoltaic is promoted by pumped storage is proposed. Based on the extracted typical scenarios of wind-photovoltaic load, a planning method of the optimal ratio is carried out. In the process, a deterministic optimization model, a robust optimization model, and a stochastic optimization model are constructed based on the wind-solar forecast errors. Among them, the most optimistic ratio is obtained in the deterministic optimization model, while the most conservative ratio is obtained in the robust optimization model.

References

1. Sang, L., Xu, Y., Long, H., Wu, W.: Safety-aware semi-end-to-end coordinated decision model for voltage regulation in active distribution network. In: IEEE Transactions on Smart Grid (2022)
2. Ge, P., Zhu, Y., Green, T.C., Teng, F.: Resilient secondary voltage control of islanded microgrids: an ESKBF-based distributed fast terminal sliding mode control approach. IEEE Trans. Power Syst. **36**(2), 1059–1070 (2021)
3. Wei, L., Weijia, Z., Jin, P., Qian, N.: Optimal capacity allocation of large-scale wind-PV-battery hybrid system. In: 2015 7th International Conference on Intelligent Human-Machine Systems and Cybernetics, pp. 420–423 (2015)
4. Kaabeche, A., Belhamel, M., Ibtiouen, R.: Sizing optimization of grid-independent hybrid photovoltaic/wind power generation system. Energy. **36**(2), 1214–1222 (2011)
5. Zhang, S., Zhang, G., Zhang, K.: Coordinated control strategy of wind-photovoltaic hybrid energy storage considering prediction error compensation and fluctuation suppression. In: 2021 IEEE 2nd International Conference on Information Technology, Big Data and Artificial Intelligence (ICIBA), pp. 1185–1189 (2021)
6. Xu, S., Liu, C., Su, C., Wang, C.: Correlation analysis of wind and photovoltaic power based on mixed copula theory and its application into optimum capacity allocation. In: 2019 IEEE 3rd Conference on Energy Internet and Energy System Integration (EI2), pp. 976–980 (2019)
7. Fang, F., Zhu, Z., Jin, S., Hu, S.: Two-layer game theoretic microgrid capacity optimization considering uncertainty of renewable energy. IEEE Syst. J. **15**(3), 4260–4271 (2021)
8. Ma, R., Li, X., Wu, Z., Zhang, Q.: Multi-objective optimal scheduling of power system considering the coordinated operation of photovoltaic-wind-pumped storage hybrid power. In: 2015 5th International Conference on Electric Utility Deregulation and Restructuring and Power Technologies (DRPT), pp. 693–698 (2015)
9. Tang, Y., Fang, G., Tan, Q., Wen, X., Lei, X., Ding, Z.: Optimizing the sizes of wind and photovoltaic power plants integrated into a hydropower station based on power output complementarity. Energy Convers. Manag. **206**, 112465 (2020)
10. MoghaddasTafreshi, S.M., Hakimi, S.M.: Optimal sizing of a stand-alone hybrid power system via particle swarm optimization (PSO). In: 2007 International Power Engineering Conference (IPEC 2007), pp. 960–965 (2007)

11. Hong, Y.-Y., Lian, R.-C.: Optimal sizing of hybrid wind/PV/diesel generation in a stand-alone power system using Markov-based genetic algorithm. IEEE Trans. Power Deliv. **27**(2), 640–647 (2012)
12. Atwa, Y.M., El-Saadany, E.F., Salama, M.M.A., Seethapathy, R.: Optimal renewable resources mix for distribution system energy loss minimization. IEEE Trans. Power Syst. **25**(1), 360–370 (2010)
13. Katsigiannis, Y.A., Georgilakis, P.S., Karapidakis, E.S.: Hybrid simulated annealing–Tabu search method for optimal sizing of autonomous power systems with renewables. IEEE Trans. Sustain.Energy. **3**(3), 330–338 (2012)

Part II
Building Energy Efficiency, Energy Management and Thermal Comfort

Chapter 7
Timber Houses in the Mediterranean Area: A Challenge to Face

Giuseppina Ciulla, Tancredi Testasecca, Stefano Mangione, Sonia Longo, and Laura Tupenaite

7.1 Introduction

Improving the energy performance of buildings and increasing the share of renewable energy is one of the three main principles on which the European Green Deal is based on [1]. As a matter of fact, a "Renovation Wave" is planned to decarbonize heating and cooling and improve the performance of public and private buildings [2]. In the European Parliament resolution of the 15th of January 2020 [3], the EU encourages the use of timber and other environmentally sustainable materials in the building sector; indeed 80% of all the international works published in the literature deal to timber construction in Europe [4]. Application of timber buildings in different European climate is a hot topic in literature. For example, Stazi et al. [5] verify the performance of a timber residential building located in a hot dry summer climate, and results showed that, in a hot climate, indoor overheating may occur without smart ventilation strategies, or the interior wall mass is too low. Kosonen et al. in [6] demonstrate how in Finland, a house made entirely of logs, without any insulation, could satisfy the definition of zero-energy building (ZEB), thanks to a correct design of the vertical surface and the generation plant. According to Svajlenka et al. [7], the modern methods of timber construction led to indisputable ecological and environmental benefits, using panel wood-based system built off-site. Since timber is a key point on new building materials, its thermophysical qualities must be analyzed, because they contribute to the building needs. The review on thermophysical properties presented by Asdrubali et al. [8] shows that timber is

G. Ciulla · T. Testasecca (✉) · S. Mangione · S. Longo
Università degli studi di Palermo, Palermo, Italy
e-mail: tancredi.testasecca@unipa.it

L. Tupenaite
Vilnius Gediminas Technical University, Vilnius, Lithuania

X. Wang (ed.), *Future Energy*, Green Energy and Technology,
https://doi.org/10.1007/978-3-031-33906-6_7

competitive in terms of thermal and structural behavior, but as illustrated in [9, 10], the thermal response of wooden walls is highly dependent on the moisture content inside them.

To evaluate how a wooden house can also represent an alternative solution to traditional constructions in a Mediterranean context, the following work proposes a case study in which the performance of a wooden building is compared with that of a building constructed with techniques traditional. In detail, thanks to an analytical model developed in MATLAB [11], the main static and dynamic thermophysical characteristics that affect the thermal balance of the building system will be highlighted. For each model, thanks to a parametric analysis, the main thermophysical and geometric characteristics necessary to achieve the minimum environmental and energy comfort requirements will be identified.

This work and research is organized as follows: In the first section, the approach and the physical variables that will be taken into consideration are described; the description of the case study follows, i.e., a traditional residential building typical of the Mediterranean area and the design of the same in timber. These buildings were simulated in different climatic zones, in order to highlight how different climatic conditions influence the design and therefore the performance of a building. Finally, the results compare the energy and economic performance between a traditional and a timber building, where the main results and contents of this work are described in the conclusions.

7.2 Materials and Methods

Based on the above, the following work proposes an analytical method for the analysis of the performance of timber houses. To be able to identify the best solution to varying climatic conditions, a case study is proposed in which the main thermophysical and dynamic parameters that come into play are evaluated. This approach will allow us to evaluate how a wooden house, unlike a traditional building with a massive structure, can, with a correct design, respond to the needs of indoor comfort even for buildings built in a Mediterranean climatic context. The following work is set up as follows: After identifying a base model, this was designed according to the actual law and simulated for different climate zones and for both the construction techniques. All models are simulated in steady state to calculate the energy consumption of the buildings and have allowed to evaluate energy consumption and possible economic savings. To complete the study, the behavior of thermophysical characteristics of the walls was studied when one of the fundamental parameters in the energy balance of a building, namely, the thickness of the insulation, varies.

7.2.1 *Building Energy Balance and Simple Economic Evaluation*

In the design of a building, the engineer must evaluate all the exchanges that the thermodynamic system of the building exchanges with the surrounding environment. The goal is to create a comfortable building for the occupants. The building is not an object, but part of an interactive and dynamic system that considers different aspects: natural (earth, water, wind, sun, vegetation), social (identity and belonging to places), technical (materials, elements), and geometric (position with respect to the sun, size of the glazed surface). Generally, energy requirements of a building are influenced by the transmission losses through the envelope, the energy gains due to solar radiation, and the presence of people and strongly correlated to the climate context and the thermophysical parameters. About the thermophysical parameters, it is important to underline how the thermal transmittance and the dynamic thermal transmittance of the envelope are fundamental in the evaluation of transmission losses and in the choice of the best air conditioning system both in summer and in winter [12]. Simultaneously the weather is one of the main factors to consider when designing a building because it represents the most important boundary condition that affects the behavior of a building.

Closely linked to the energy needs of a building are the costs related to the use of the heating and/or cooling system and the production of DHW. If a building requires the consumption of electricity only to achieve indoor comfort, it is possible to convert $PE_{g,\,nr}$, the global primary energy non-produced by local renewable energy sources (RES) required by the building, into an annual operating cost A, according to Eq. (7.1):

$$A = \frac{PE_{g,nr}}{f_{p,\text{tot}}} \bullet S_{\text{floor}} \bullet E_{e,\text{price}} \tag{7.1}$$

where S_{floor} is the floor building surface [m^2]; $f_{p,\,\text{tot}}$ is the total primary energy conversion factor, which is 2.42 for Italy [13]; and $E_{e,\,\text{price}}$ is the price assumed for 1 kWh of electric energy [€/kWh], which is assumed 39.6 c€/kWh as the mean price of 2022 in Italy [14]. Considering different interest rate *i*, the net present value (NPV) and the discounted payback time (DPT) can be calculated.

7.2.2 *Thermophysical Qualities*

Since the building envelope is responsible for the thermal heating and cooling energy needs, it is important to deepen the knowledge of the properties of building materials, analyzing how the quantity or geometry of these affect the quality of the building envelope. For this type of analysis, a code has been developed in a MATLAB environment. For a given stratigraphy of the wall, depending on the climatic zone, the model generates a curve that allows to identify the thickness of the insulation necessary to meet the legal limits in terms of stationary thermal transmittance U, periodic thermal transmittance Y, surface of the mass M_S, and displacement time Δt. This analysis was developed for both classic brick and cross-laminated timber (CLT), keeping the thickness of the load-bearing structure constant.

7.3 Case Study

To highlight how energy consumption is closely linked to the type of building that is designed and used, the following paper analyzes the energy performance based on the following steps:

1. The performance of a residential building designed according to the traditional typology of masonry houses located in the south of Italy will be analyzed.
2. The same building will be redesigned in wood, to evaluate how this type of building has a great benefit on consumption and therefore on the environment.
3. To assess how the climate plays an important role in these assessments, the performance of the same building at different latitudes and climate context will be analyzed and compared.
4. To assess how the thermophysical parameters are fundamental in the energy and economic evaluation, a focus on the envelope quality will be described.
5. Analysis of the results: consumption analysis and economic investigation.

7.3.1 Brick House

It is assumed to consider a residential house of simple geometry, built with traditional materials, that is with reinforced concrete structure, external walls in perforated bricks, wood frames and double glazing, insulated attic with simple sheath, with an uninhabited attic and ending with a double pitched roof with tiles (see Fig. 7.1).

The heating system is composed by a hydronic heat pump and radiators, and the cooling system is composed by a split system. The domestic hot water is provided by heat pump and flat plate collectors. A thermal storage is installed too to increase the efficiency of the system. To achieve the maximum rate of renewable energy, there is also a photovoltaic system (PV) of 5.6 kWp and a battery with a capacity of 9.8 kWh. The energy performance of this house was evaluated in six different cities of the Southern Italy; Table 7.1 indicates the six cities, the heating degree day (HDD) values, and the thermal transmittance limits of the envelope, respectively.

According to the Italian national guidelines for buildings energy certification, it was possible to identify different climatic zones that (theoretically) have the same

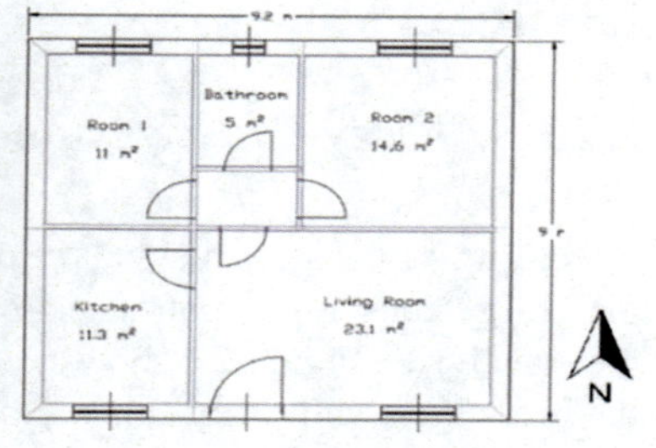

Fig. 7.1 3D render and schema of the building

Table 7.1 Italian cities and parameters

Cities	Porto Empedocle	Palermo	Ragusa	Caltanissetta	Enna	Floresta
Climatic zone	A	B	C	D	E	F
HDD [°C/day]	579	751	1050	1550	2248	3309
$U_{\text{limit, wall}}$ [W/m^{2}K]	0.4	0.4	0.36	0.32	0.28	0.26
$U_{\text{limit, roof}}$ [W/m^{2}K]	0.32	0.32	0.32	0.26	0.24	0.22
$U_{\text{limit, floor}}$ [W/m^{2}K]	0.42	0.42	0.38	0.32	0.29	0.28
$U_{\text{limit, window}}$ [W/m^{2}K]	3	3	2	1.8	1.4	1

climate [15]. Employing the HDD, it is possible to identify six different climatic zones: zone A represents the hottest one and zone F the coldest. In each location it was implemented a house with the envelope that respect the requirements indicated from the Italian legislation [13]. More in detail, all six models (base case) have the same construction; the only difference is the insulation thickness that changes for each climate zone, while the same composition was chosen for the roof and the floor, satisfying the legislative limits for the coldest city, Floresta.

7.3.2 *Timber House*

To evaluate how the same building, made of wood, allows for a reduction in consumption, leaving the geometry and position with respect to the sun unchanged, it was assumed that the external walls are made of CLT with the presence of rock wool panels. In the "Timber Case", six models were developed for CLT structures, one for each climatic zone. The thickness of the walls insulation layer was adjusted for each zone and the same floor and roof were used to meet the transmittance limit of the F climatic zone. More in detail, the floor is designed with CLT, a vapor barrier, and a cork panel of 15 cm. The roof structure is like the brick one, but the slab is substituted with wood beams and the insulation chosen is a recycled extruded polystyrene (XPS).

7.4 Results

7.4.1 Energy Consumption and Comparison

According to the Italian standard procedure and using a dedicated software [16], the energy consumption of the six base case models and the six timber case models were analyzed. Figure 7.2 shows the energy needs of all simulated models. For both the base case and the timber case, the maximum primary energy requirement occurs in Caltanissetta, while the minimum is reached in the city of Porto Empedocle. Obviously, the energy required for the heating period decreases for less rigid climates, while the need for cooling increases. An evaluation of the energy savings related to wooden houses shows a reduction that varies from 17.19% for zone F to 0.23% in zone B. In the case in which the thermal energy required is very low, such as the climatic zone A, the global primary energy is higher of a factor 1.038. As explained in Fig. 7.2, in all cities, the cooling energy required of the timber house is higher than the traditional ones. For instance, in Porto Empedocle PE_h passes from 9.6 kWh/m^2 for the base case to 5.76 kWh/m^2 of the timber case, while the PE_c increases of 6.2 kWh/m^2.

The production of energy for heating and cooling is partially satisfied by the RES on site. The photovoltaic panels power the heat pumps, while the solar collectors integrate the production of DHW. The right chart of Fig. 7.2 shows that the non-renewable energy (NREN) required is lower in the timber house positioned in warm climates. In fact, in zones A and B, there is no NREN energy consumption, (PE_{nr}). It is fundamental to highlight that, in all cases, the cooling energy requirement is completely satisfied by the PV system. The reduction of PE_{nr} in the timber houses is between 31% (zone F) and 100% (zone B) so, in general, timber buildings require less PE_{nr} then base case.

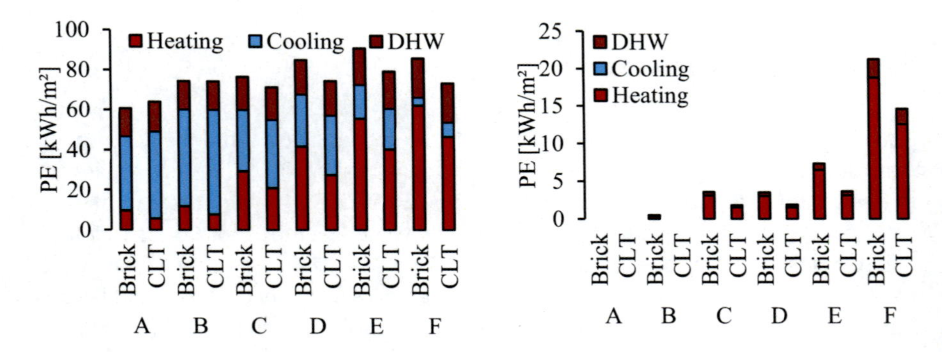

Fig. 7.2 Primary energy consumption (left) and primary energy not produced by RES (right)

7.4.2 Economic Investigation

In general, to date wooden constructions require higher prices based on [17] and in general require about 3–8 k€ or more. In the coldest cities (Floresta and Enna), the brick walls require a thickness of insulation of 8 cm greater than timber and these are the only cases in which the CLT walls cost 3600 € less. Considering the insulation, costs are generally lower in wooden constructions due to the lower thermal transmittance value, thus requiring a reduced thickness of insulation. As indicated in Table 7.2, the yearly costs are reduced by a maximum of 100% in Zone B and a minimum of 33.6% in zone E.

As can be seen in Fig. 7.3, the differences between the different annual savings and lumber costs, spread over 20 and 40 years, for each model, are considerable. Without considering any interest rate, the two configurations in which the timber is economically more sustainable are in zone E and in zone F. For the climatic zones A–D, the investment costs are the same, because in each model the difference between the thickness of the insulation, from the base case to the timber case, is always the same.

For the climatic zones E and F, a sensitivity analysis was carried out for different interest rates. The results show that using wood, for buildings built in coldest city, will take 20 years to pay back the investment considering an interest rate of 10%. With lower rates, the DPT drops dramatically and becomes just 13 years for a 6% interest rate and only 10 years for an assumed 3% interest rate. In this case, after 40 years, the NPV will be greater than 6000 €. For Enna, the city that represents the climate zone E, timber house, with an interest rate less than 3%, will be characterized by a NPV after 50 years of only 2000 €, with a DPT that exceeds 20 years.

Table 7.2 Yearly costs for electric energy for each model

Climatic zone	A	B	C	D	E	F
Base case [€/y]	0	30.15	214.7	211.7	442.6	884.7
Timber case [€/y]	0	0	109.8	140.5	220.7	397.4

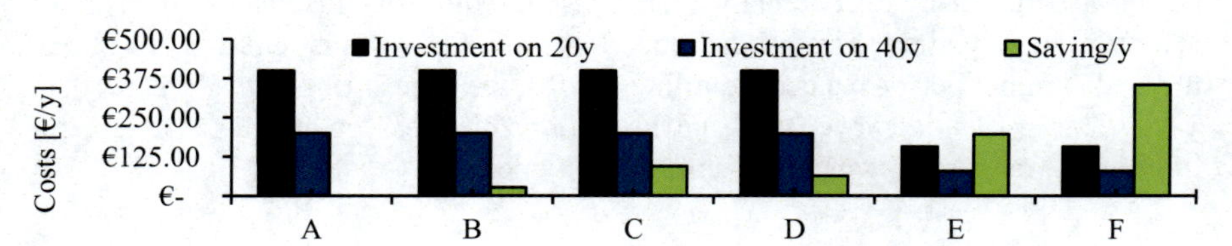

Fig. 7.3 Investment costs and saving per year

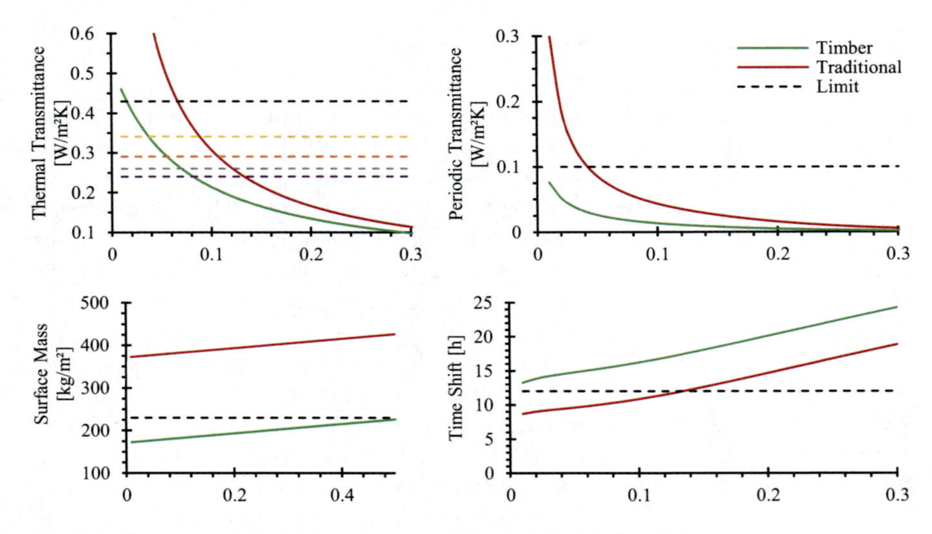

Fig. 7.4 Thermophysical variables trend as a function of insulation thickness

7.4.3 *Thermophysical Qualities*

The study on the thermophysical quality of the envelope is represented in Fig. 7.4, in which the value of the transmittance is represented as a function of the thickness of the insulation for the brick building and for the CLT building, respectively. The horizontal lines refer to the legal limit values for the various areas. The results show that regardless of the climatic zone, CLT panels require less insulation because they are characterized by a lower thermal conductivity than brick ones. Wooden walls require an insulation thickness of about 5–6 cm less [13]. As can be seen in Fig. 7.4, the traditional walls satisfy the superficial mass requirement without any insulation, while for CLT panels, an insulation thickness greater than 50 cm is required. This happens because both CLT and insulation are characterized by low density. In the evaluation of periodic thermal transmittance, the results show that the CLT wall still performs better than the concrete wall, because it requires an insulation thickness of less than 1 cm, while for traditional one, a thickness of 4 cm is required. To obtain the best summer performance, according to [18], the time shift must be greater than 12 h. This goal is achieved with an insulation thickness greater than 13 cm for traditional walls and lower than 1 cm for CLT ones.

7.5 Conclusions

In this work the authors propose the use of timber houses in the Mediterranean area. This type of building is generally typical of cold climates and Northern Europe, but to meet the needs of reducing consumption and costs, the first results of a case study of houses in wood located in Southern Italy compared to a traditional building are described. Since the climate is an important factor in evaluating the energy consumption of buildings, authors propose the study of the energy consumption of these buildings in six different climatic zones. Furthermore, to evaluate how the energy quality is dependent on a correct design and identification of some furnishing parameters, the fundamental thermal properties have been analyzed with a code implemented on MATLAB environment. At the same time, a simple economic analysis was carried allowing to evaluate the cost of the investment. The analysis of the first results shows that wooden buildings obtain energy savings ranging from 17.19% to 0.23%, depending on the climatic zone, except the hottest city. Wooden constructions are economically competitive in colder climates, such as Floresta and Caltanissetta. Furthermore, since solar radiation is high in Sicily, the increased demand for cooling in timber buildings could be fully covered by the presence of a PV plant. The model developed on MATLAB showed how a timber house requires less insulation to achieve the standards of periodic and stationary transmittance and time shift. In conclusion, the research work presented shows that wooden buildings can be attractive solutions even in the Mediterranean climate bringing environmental benefits and cost savings. Despite that, currently, timber buildings are still not economically competitive compared to traditional constructions. This work is only the first result of a research that will be deepened and refined, to represent a guide for the new-generation designers who must face complex challenges that consider both energy, environmental, and economic aspects at the same time.

Acknowledgments This research was supported by the EU Erasmus+ project "Circular Economy in Wooden Construction" (Wood in Circle). Project No: KA203-8443DA0D. Project code: 2020-1-LT01-KA203-077939.

References

1. European Commission: Una transizione all'energia pulita 2020
2. European Commission: Building and renovating: the European Green Deal, No. December, pp. 1–10 (2019). https://doi.org/10.2775/48978
3. European Parliament: Resolution of the European Parliament of 15-01-20 (2020)
4. Caniato, M., Marzi, A., da Silva, S.M., Gasparella, A.: A review of the thermal and acoustic properties of materials for timber building construction. J. Build. Eng. **43**(August), 103066 (2021)
5. Stazi, F., Tomassoni, E., Bonfigli, C., di Perna, C.: Energy, comfort and environmental assessment of different building envelope techniques in a Mediterranean climate with a hot dry summer. Appl. Energy. **134**, 176–196 (2014)
6. Kosonen, A., Keskisaari, A.: Zero-energy log house – future concept for an energy efficient building in the Nordic conditions. Energy Build. **228**, 110449 (2020)
7. Švajlenka, J., Kozlovská, M.: Evaluation of the efficiency and sustainability of timber-based construction. J. Clean. Prod. **259**, 120835 (2020)
8. Asdrubali, F., Ferracuti, B., Lombardi, L., Guattari, C., Evangelisti, L., Grazieschi, G.: A review of structural, thermo-physical, acoustical, and environmental properties of wooden materials for building applications. Build. Environ. **114**, 307–332 (2017)

9. Stazi, F., Ulpiani, G., Pergolini, M., di Perna, C.: The role of areal heat capacity and decrement factor in case of hyper insulated buildings: an experimental study. Energy Build. **176**, 310–324 (2018)
10. Bienvenido-Huertas, D., Rubio-Bellido, C., Pulido-Arcas, J.A., Pérez-Fargallo, A.: Towards the implementation of periodic thermal transmittance in Spanish building energy regulation. J. Build. Eng. **31**(December), 2020 (2019)
11. The Math Works Inc.: MATLAB. Version 2020a. The Math Works, Inc. (2020). Accessed 14 Oct 2022. [Online]. Available: https://www.mathworks.com/
12. Ciulla, G., Lo Brano, V., D'Amico, A.: Numerical assessment of heating energy demand for office buildings in Italy. Energy Procedia. **101**, 224–231 (2016)
13. Economico, M.D.S.: Decreto Interministeriale 26 Giugno 2015- Economic: Rome, Italy (2015)
14. ARERA: Andamento del prezzo dell'energia elettrica per il consumatore domestico tipo in maggior tutela (2022). https://www.arera.it/it/dati/eep35.htm. Accessed 14 Oct 2022
15. Decreto del Presidente della Repubblica *26 agosto 1993, n. 412. Available online: https://www.normativa. it/uri-res/N2Ls*
16. Logical Soft: TERMOLOG 12 (2021). Accessed 14 Oct 2022. [Online]. Available: https://www.logical.it/software-per-la-termotecnica
17. Provinciale, A., Le, P., Pubbliche, O.: Provincia autonoma di trento dipartimento infrastrutture (2022)
18. Decreto Ministeriale 26/6/2009-Ministero dello Sviluppo Economico Linee guida nazionali per la certificazione energetica degli edifici

Chapter 8
Internet of Things-Based Smart Building for Energy Efficiency

Muhammad R. Ahmed, Thirein Myo, Mohammed A. Aseeri, Badar Al Baroomi, M. S. Kaiser, and Woshan Srimal

8.1 Introduction

The environment sustainability requires minimizing human activities' impacts on the region in which they take place. Over the next few years, energy demand will increase significantly due to population growth, economic growth, and consumer demand for more comfortable life [1].

A large portion of the world's energy demand comes from buildings. As a result, it appears that in order to meet the world's sustainability challenge, the energy efficiency of buildings needs to be enhanced [2]. Considering this, there has been a shift in the way buildings are being utilized beyond just constructions that provide shelter. Internet of Things (IoT) gave a new dimension to the buildings and make the building as smart building [3]. A building can be defined as smart building that has processes for automating its operations, such as heating, air conditioning, ventilation, electricity management, security systems, and more [4]. Creating a smart building allows us to reduce overhead and resolve issues in order to avoid them in the future. An efficient and secure building contributes to the health of the environment and the productivity of its occupants. The core of smart buildings is the integration of water, power, and security with optimal uses. Buildings equipped with IoT technology offer the possibility of monitoring optimal uses of energy and

M. R. Ahmed (✉) · T. Myo · B. Al Baroomi · W. Srimal
Military Technological College, Muscat, Oman
e-mail: muhammad.ahmed@mtc.edu.om

M. A. Aseeri
King Abdulaziz City for Science and Technology (KACST), Riyadh, Saudi Arabia
e-mail: masseri@kacst.edu.sa

M. S. Kaiser
Institute of Information Technology, Jahangirnagar University, Savar, Bangladesh

X. Wang (ed.), *Future Energy*, Green Energy and Technology,
https://doi.org/10.1007/978-3-031-33906-6_8

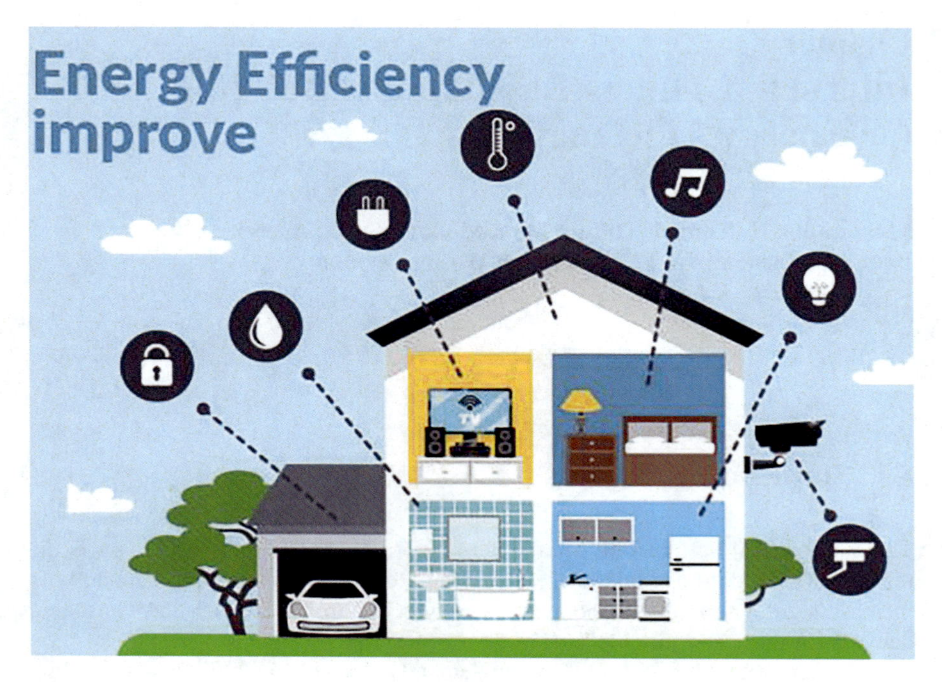

Fig. 8.1 The energy-efficient building

enhancing efficiency [5]. IoT sensors are used to collect the data. Analyzing this data allows us to decide what resources to allocate based on the analysis of the data [6]. The highest loads should be prioritized before the lowest loads for each monitoring device. Figure 8.1 shows the energy-efficient building equipped with different sensors and actuators.

Incorporating IoT into buildings has many benefits, including increased energy efficiency in the buildings. The Internet of Things is composed of sensors, actuators, cloud-based software, and communication protocols [7]. In the building, to help optimize the systems, they are layered and controlled by a central building management system (BMS), which enables them to communicate with each other and work together. Having sensors in buildings allows them to collect information about the ambient conditions inside the building and available resources [8]. We can collect a wide range of data using sensors, such as temperature, humidity, light intensity, airflow, and smart energy meters, among others. Actuators include anything that can be controlled, such as light switches, windows, lifts, doors, air conditioners, ventilation systems, and presence detectors, among others [9]. Interconnecting these elements requires information and communication technology. To make the actuators respond to control unit commands, the sensors send information directly to the control unit, which then passes the commands directly on to the actuators. Information technology was incorporated to efficiently process and store the information collected from the building's sensors [10]. The data is stored in a

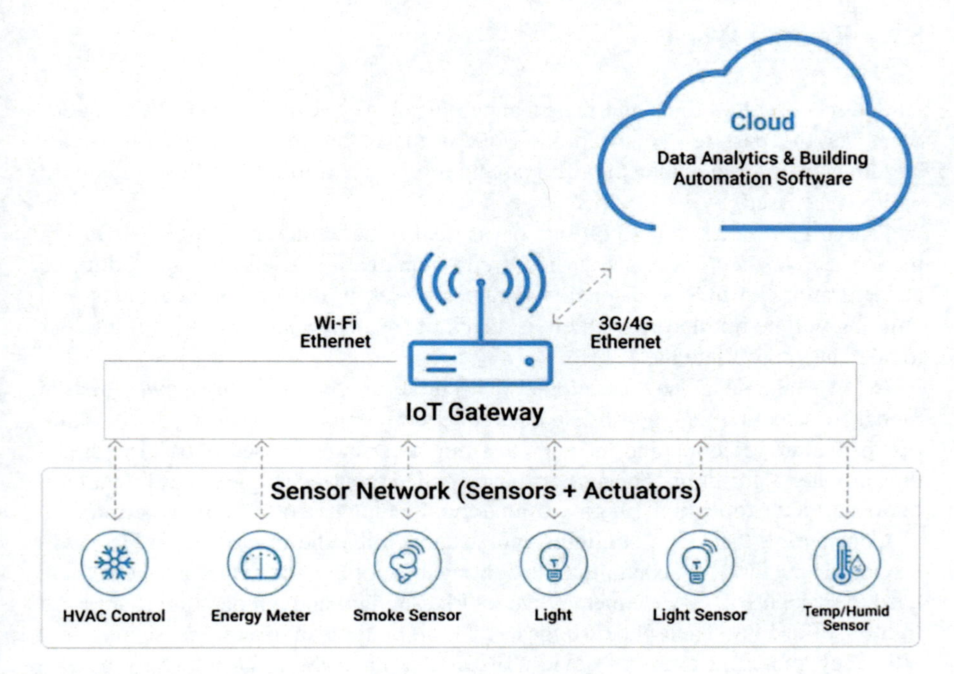

Fig. 8.2 The IoT sensors and actuators

database for later retrieval and use. In addition to storing data, smart buildings use modern analytics techniques to provide an in-depth understanding of a building's energy consumption. The results of analyses can be used to predict peak and valley times, the energy consumption of different equipment over time, and the use of appliances [11]. Using this analytical information, the individual can optimize their comfort and energy consumption based on current data, historical data, as well as evolution of their power consumption. Figure 8.2 portrayed the use of the sensors and actuators with data analytics and automation.

A building consumes power for its various equipment and systems. These include heating systems, air conditioning systems, hot water systems, lighting, and commodities. By utilizing IoT devices and building management systems, smart buildings can manage [12]:

- The energy by using smart meters, demand-response systems, and other energy-saving systems
- Elevators, lighting, and other equipment by utilizing daylight meters, presence sensors, lift demand sensors, etc.
- Lighting, HVAC, and windows to comfort in the environment

In the remaining paper, the following structure is followed: The overview of some recent existing works is discussed in Sect. 8.2, followed by a discussion of the architecture of the Internet of Things in Sect. 8.3. We described the Internet of Things-based smart buildings in Sect. 8.4 and the conclusion in Sect. 8.5.

8.2 Related Work

Although smart buildings and their technologies have received considerable attention over the past few years, more work needs to be done in IoT-based smart buildings in order to realize their full potential for energy efficiency. Some literature will be discussed.

Kumar et al. [13] worked on indoor air quality of residential buildings. During their research, they developed an IAQ sensor that can be used for monitoring the concentrations of indoor air quality parameters. Using this sensor, occupants of a building will be notified when there is a lack of fresh air inside a building, in order to take the appropriate measures.

In [14] Haider et al. have developed an accurate and precise occupancy prediction model by selecting the appropriate data collection period and sensors. A combination of feature selection and machine learning algorithms is used in order to create the classifier algorithms. The authors have applied their techniques to different time periods of data collection, ranging from a minute long to a 60-min long period.

Considering the green buildings and sustainability, the researches in [15] have proposed a method for scheduling the lighting technology. In the research, LEDs are used to make use of environmental factors like sunlight and temperature to improve lighting efficiency. Their paper suggests that this could lead to an increase in energy efficiency and a clustering of devices distributed throughout the house for a more intelligent and autonomous control.

In [16] Alexander et al. proposed a data analysis system for the temperature. Temperature sensors can be used by the application to gather detailed information about the temperature of the environment. In the proposed approach, it is intended to lay the groundwork for the development of an intellectual data analysis system. The system is capable of modeling the temperature mode and controlling the heating process of a building based on the model of the temperature mode. To do that they used three-layer based model.

Combining heat pumps with photovoltaic (PV) generators provides both electric and thermal loads, which are converted into electrical energy by utilizing the energy produced by the PV generator that was presented in [17]. In designing this system, the researchers have considered that it is totally self-sufficient and that it relies as little as possible on the grid. In order to achieve the optimal compromise between costs and benefits, they have been conducting energy simulations on a regular basis, in order to evaluate the size of the batteries on a yearly basis.

Khan et al. have proposed a fuzzy logic-based model for energy efficiency in [18]. With the aim of reducing the consumption of energy in buildings by using the central heating system, they designed a control system and simulated it in buildings. With the help of artificial intelligence, the proposed model reduces the energy consumption in buildings by improving the quality of life and comfort of the occupants.

Ali et al. proposed a thermal control for air conditioning systems using IoT in [19]. Their work is based on an Internet of Things (IoT)-based smart system, which, according to the researchers, allows them to control the AC to provide a thermally comfortable environment. By recording the user's feelings toward the environment as input for the system, their created system will be able to interact with the users by interacting with the system. Using the sensor data together with an enhanced predicted mean vote (PMV)-based model, this information can be integrated together with the AC control system so that the AC can be controlled intelligently so that the occupant is able to experience thermal satisfaction.

In [20] Zhang et al. presented a method that utilizes the low-resolution temperature readings that are obtained from IoT-based smart thermostats to create a thermal model. A building's Internet of Things platform was used to collect data over a summer period to validate the learning framework. In addition, the researchers have evaluated and quantified the accuracy of indoor temperature prediction based on the learned model as well as the performance of the learned model.

Rastegarpour et al. [21] presented a study and developed a novel control method for integrating thermal energy storage (TES) systems, HVAC systems, buildings, and local renewable energy sources in order to be used in conjunction with optimization techniques for building energy management. Through model predictive control (MPC), the control framework is able to predict the effects of disruptions. As an additional benefit, the novel configuration of the TES in combination with a heat pump and a radiant floor building makes for a complex model.

The demand for energy in buildings is increasing due to the growing population, the need to reduce carbon footprint, and the smart grid paradigm, making IoT-based smart buildings essential. The work found in the literature focuses on the specific energy consumption and efficiency. This leads for a need to have an architecture for smart building that focuses most on energy consumption in the building.

8.3 Architecture of IoT

Internet of Things refers to a network of interconnected devices, such as digital technology and mechanical devices, which are interconnected in a way that does not require human interaction. A system based on the Internet of Things consists of web-enhanced devices, processors, communication hardware, and sensors, which are responsible for sending and deploying data to provide the desired outcomes.

IoT architectures integrate various components that work together harmoniously to accomplish a specific task. The Internet of Things will function differently based on how it is designed and developed [22]. In an Internet of Things architecture, data are collected from sensors; devices are operated; information is sent or received; data is sensed, stored, processed, and analyzed; cloud and edge-based services are provided; applications are provided; and data is exploited by end users [23]. The Internet of Things will work differently in different application areas. IoT frameworks are used to implement a wide range of technologies based on the application. There is, however, not a standard defined architecture for how the IoT

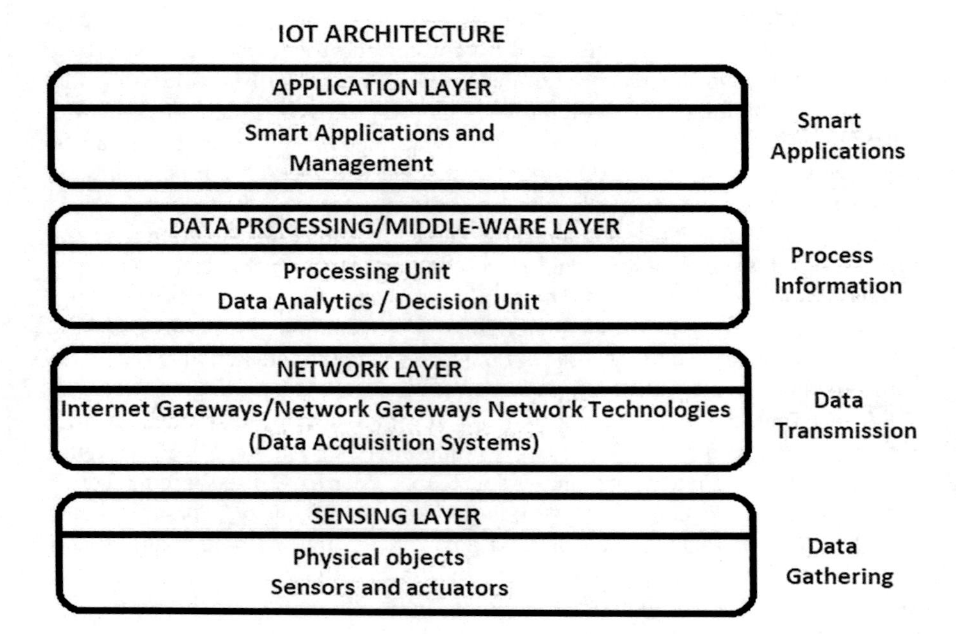

Fig. 8.3 The IoT four-layer architecture

will work, which is strictly followed universally. In the literature, it has been found that the common IoT architecture consists of three or four layers. In their layer architecture, the major layers are application layer, network layer, and physical layer. In the four-layer architecture, it is sensing layer, network layer, middleware layer/data processing layer, and application layer [24]. The four-layer architecture is shown in Fig. 8.3.

- Sensing layer: This layer of sensing consists of sensors, actuators, devices, and other components. Various sensors and actuators receive data (physical or environmental parameters), process the data, and send the data as needed.
- Network layer: Several gateways and data acquisition systems (DASs) are present in this layer, including network gateways. Data agglomeration and conversion is the function of a DAS; that is, it collects, aggregates, and converts the analog data from sensors into digital data, by aggregating and converting the data. A gateway's primary function is to enable communication between sensor networks and the Internet. Apart from these basic functions, gateways also provide malware protection, filtering, and the ability to make decisions based on inputted data, as well as data management.
- Data processing layer: The IoT processing layer/middleware layer is the heart of the IoT ecosystem. Data is collected, analyzed, and preprocessed here before being sent to a processing center. This is the place where a software application can access the data, and it's here that the data can be monitored and managed

and where further actions can be prepared. As a result, edge computing or edge analytics comes here.

- Application layer: The IoT architecture has four layers, of which this is the last layer in the IoT architecture. In this layer, the user has access to various facilities in order to satisfy individual requirements. There are a variety of applications that are incorporated into this layer.

Moreover, a variety of realistic communication models can be adopted in IoT applications depending on their specific characteristics, all of which have their own advantages. The communication model includes [25]:

- A node in the Internet of Things is connected to a device.
- A node in the IoT network is connected to a cloud.
- A node is connected to a base station for IoT.

Sensor boards are devices that contain sensors for detecting and responding to inputs generated by the physical environment. An IoT device typically contains several sensors, microcontrollers (MCUs), Bluetooth or Wi-Fi radios, and power management. In addition, central processing units come with wireless chips and components [26]. The most common IoT boards available in the market are Arduino nano, Arduino uno, Intel Galileo, and MICAz.

8.4 IoT in Smart Buildings

The use of smart devices is becoming ubiquitous in our lives, ranging from sensors to household appliances to smartphones. IoT-based networks can be consolidated by integrating smart elements in this way, resulting in heterogeneous networks. It is an integrated process that encompasses the entire lifecycle of the building and is an integral part of optimizing energy efficiency using IoT in the building [27]. In order to optimize the main stages:

- An energy-efficient design is created by using simulations to predict the energy performance.
- A subsystem is constructed and tested separately as part of the construction process.
- Monitoring and controlling the building's operation, as well as controlling the actuators.
- Maintaining infrastructure, resolving problems associated with infrastructure because of energy deficiencies.
- The process of demolition, recycling materials, and repurposing the usable elements.
- The occupancy rate of the buildings

Through the use of IoT, energy efficiency can be enhanced through automation. Data gathered by smart sensors can be used to control buildings for a number of

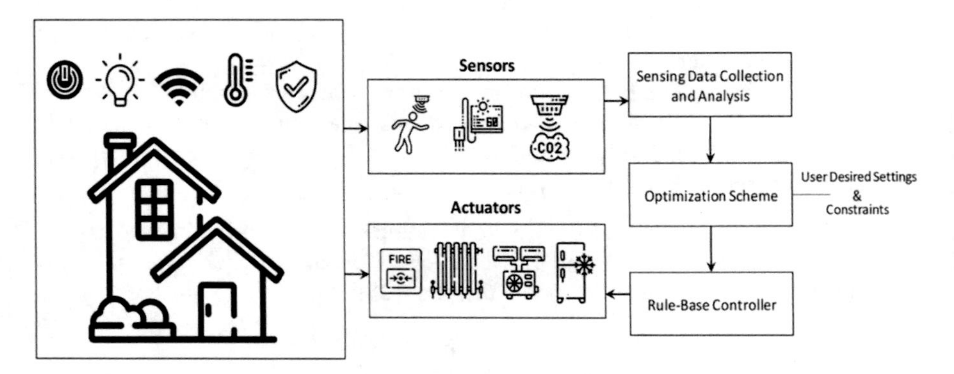

Fig. 8.4 The IoT sensors, actuators, and optimization of energy use

purposes, including turning off lights in unoccupied rooms, reducing airflow in buildings on weekends, or automatically closing blinds when windows are exposed to direct sunlight. A model of optimized energy uses is shown in the figure. The major sensor and actuators used to make the building energy efficient are daylight sensor, HVAC controllers, occupancy sensors, smart thermostats, and smart meters [28]. Figure 8.4 shows the energy optimization based on the user setting and collected data from the sensors.

- Daylight sensors: Using photocells, these devices adjust the lighting to match the amount of natural light in the area. There is also the possibility of using these photosensors to raise or lower blinds in order to improve the quality of the lighting. In most cases, they are able to save good amount of energy.
- HVAC controllers: The use of IoT-based HVAC controllers could result in a reduction in energy. Sensors are used within a building to monitor various conditions. This data is used to regulate outputs within the climate control systems. In a smart control system, predefined set points are typically used to determine what actions should be taken and they can be positioned at certain key points or throughout the building. As technology advances, some sensors and controls have begun to implement machine learning (ML) algorithms that are able to implement real time.
- Occupancy sensors: It is possible to use infrared or ultrasonic sensors to control the lighting in a room based on whether or not it is occupied. Such sensors can also control the heating and cooling of the room as well. The use of occupancy sensors is able to generate savings of the energy when it comes to lighting systems.
- Smart thermostats: This technology allows for the remote monitoring and control of the temperature of built environments. Advanced models provide reports on the amount of energy a home is saving on a monthly basis, as well as a number of analytics about energy usage.

- Smart meters: Energy efficiency in buildings is dependent upon a number of factors, including the Internet of Things, including metering and monitoring. In order to track energy consumption, smart meters are an essential tool. A smart electric submeter can also be used to track the energy consumption of individual plugs. Further, submeters are useful in identifying anomalies that may indicate that there is some kind of maintenance issue affecting energy consumption. As a result it helps to save the energy.
- Variable speed devices: It is important to use equipment's equipped with the variable speed devices. Inverter motor drives, variable frequency drives, and adjustable speed drives are also called variable frequency drives. Older buildings typically use a single-speed HVAC system, using dampers, throttles, and valves to control airflow. Since these fans operate only at full speed, the amount of energy they consume is significant as a result of this energy wastage. The variable speed drives are able to operate a fan at a higher or a lower speed, depending upon how much energy is required. There is no doubt that variable speed drives reduce energy consumption for heating and cooling in buildings.

In the Internet of Things, sensors are connected to the systems, equipment, and other infrastructure of a building to optimize building efficiency and performance and help reduce energy costs. Energy consumption reduction is only one method of promoting energy efficiency. Furthermore, the observer may also be able to prevent catastrophic system failures as well as prevent maintenance problems before they become problems. Maintenance issues can be detected using IoT sensors, which include [29]:

- Detection of early faults using sensors that can alert the user when an unseen issue signals the presence of an impending fault.
- This sensor is designed to detect a fault and if it detects a fault, the machinery will be disabled to prevent wider problems, preventing failures and downtime.

8.5 Conclusion

In the current climate change situation, green energy and sustainability become essential. As building performance requirements become increasingly demanding, buildings are being put under greater pressure in order to meet them. In order to ensure the energy sustainability of the planet, energy-efficient buildings have been recognized as an important goal. This goal has been addressed in a variety of ways, and the most recent attempt relates energy efficiency of the building with human occupancy patterns. Throughout the advancements in the field of microelectronics, efficient sensors have been developed, which has led to the development of the Internet of Things. This technology is used to make the building smart building. Due to this, the building started to be automated. Using the specific sensors and measuring and controlling the different equipment, we are able to use optimum energy and make the building more energy efficient. In this paper we have presented the IoT-based energy-efficient building with the IoT-based architecture of the building. According to the architecture, we will be able to create a building that is energy efficient with the use of IoT with optimum use of the energy. It is our intention to highlight our contribution in the implementation of real-world sensors in the building so that the energy efficiency of the building can be measured in real-time performance.

References

1. Nejat, P., Jomehzadeh, F., Taheri, M.M., Gohari, M., Majid, M.Z.A.: A global review of energy consumption, CO2 emissions and policy in the residential sector (with an overview of the top ten CO2 emitting countries). Renew. Sustain. Energy Rev. **43**, 843–862 (2015). https://doi.org/10.1016/j.rser.2014.11.066
2. Bottaccioli, L., et al.: Building energy modelling and monitoring by integration of IoT devices and building information models. In: 2017 IEEE 41st Annual Computer Software and Applications Conference (COMPSAC), vol. 1, pp. 914–922 (2017). https://doi.org/10.1109/COMPSAC.2017.75
3. Yu, L., Qin, S., Zhang, M., Shen, C., Jiang, T., Guan, X.: A review of deep reinforcement learning for smart building energy management. IEEE Internet Things J. **8**(15), 12046–12063 (2021). https://doi.org/10.1109/JIOT.2021.3078462
4. Doukari, O., Seck, B., Greenwood, D., Feng, H., Kassem, M.: Towards an interoperable approach for modelling and managing smart building data: the case of the CESI smart building demonstrator. Buildings **12**, no. 3, Art. no. 3 (2022). https://doi.org/10.3390/buildings12030362
5. Samad, T., Koch, E., Stluka, P.: Automated demand response for smart buildings and microgrids: the state of the practice and research challenges. Proc. IEEE. **104**(4), 726–744 (2016). https://doi.org/10.1109/JPROC.2016.2520639
6. Cicioğlu, M., Çalhan, A.: A multiprotocol controller deployment in SDN-based IoMT architecture. IEEE Internet Things J. **9**(21), 20833–20840 (2022). https://doi.org/10.1109/JIOT.2022.3175669
7. Sittón-Candanedo, I., Alonso, R.S., García, Ó., Muñoz, L., Rodríguez-González, S.: Edge computing, IoT and social computing in smart energy scenarios. Sensors **19**, no. 15, Art. no. 15 (2019). https://doi.org/10.3390/s19153353
8. Balaji, B., Verma, C., Narayanaswamy, B., Agarwal, Y.: Zodiac: organizing large deployment of sensors to create reusable applications for buildings. In: Proceedings of the 2nd ACM International Conference on Embedded Systems for Energy-Efficient Built Environments, pp. 13–22. New York (2015). https://doi.org/10.1145/2821650.2821674
9. Natarajan, A., Krishnasamy, V., Singh, M.: Occupancy detection and localization strategies for demand modulated appliance control in Internet of Things enabled home energy management system. Renew. Sustain. Energy Rev. **167**, 112731 (2022). https://doi.org/10.1016/j.rser.2022.112731
10. Chouhan, P.K., McClean, S., Shackleton, M.: Situation assessment to secure IoT applications. In: 2018 Fifth International Conference on Internet of Things: Systems, Management and Security, pp. 70–77 (2018). https://doi.org/10.1109/IoTSMS.2018.8554802
11. Misra, S., Roy, C., Sauter, T., Mukherjee, A., Maiti, J.: Industrial Internet of Things for safety management applications: a survey. IEEE Access. **10**, 83415–83439 (2022). https://doi.org/10.1109/ACCESS.2022.3194166
12. Shah, S.F.A., et al.: The role of machine learning and the Internet of Things in smart buildings for energy efficiency. Appl. Sci. **12**, no. 15, Art. no. 15 (2022). https://doi.org/10.3390/app12157882

13. Kumar, A., Kumar, A., Singh, A.: Energy efficient and low cost air quality sensor for smart buildings. In: 2017 3rd International Conference on Computational Intelligence & Communication Technology (CICT), pp. 1–4 (2017). https://doi.org/10.1109/CIACT.2017.7977310
14. Haidar, N., Tamani, N., Nienaber, F., Wesseling, M.T., Bouju, A., Ghamri-Doudane, Y.: Data collection period and sensor selection method for smart building occupancy prediction. In: 2019 IEEE 89th Vehicular Technology Conference (VTC2019-Spring), pp. 1–6 (2019). https://doi.org/10.1109/VTCSpring.2019.8746447
15. Jeyasheeli, P.G., Selva, J.V.J.: An IOT design for smart lighting in green buildings based on environmental factors. In: 2017 4th International Conference on Advanced Computing and Communication Systems (ICACCS), pp. 1–5 (2017). https://doi.org/10.1109/ICACCS.2017.8014559
16. Zakharov, A., Romazanov, A., Shirokikh, A., Zakharova, I.: Intellectual data analysis system of building temperature mode monitoring. In: 2019 International Russian Automation Conference (RusAutoCon), pp. 1–6 (2019). https://doi.org/10.1109/RUSAUTOCON.2019.8867611
17. Leo, P.D., Spertino, F., Fichera, S., Malgaroli, G., Ratclif, A.: Improvement of self-sufficiency for an innovative nearly zero energy building by photovoltaic generators. In: 2019 IEEE Milan PowerTech, pp. 1–6 (2019). https://doi.org/10.1109/PTC.2019.8810434
18. Ilhan, İ., Karaköse, M., Yavaş, M.: Design and simulation of intelligent central heating system for smart buildings in smart city. In: 2019 7th International Istanbul Smart Grids and Cities Congress and Fair (ICSG), pp. 233–237 (2019). https://doi.org/10.1109/SGCF.2019.8782356
19. Ali, A.M., Shukor, S.A.A., Rahim, N.A., Razlan, Z.M., Jamal, Z.A.Z., Kohlhof, K.: IoT-based smart air conditioning control for thermal comfort. In: 2019 IEEE International Conference on Automatic Control and Intelligent Systems (I2CACIS), pp. 289–294 (2019). https://doi.org/10.1109/I2CACIS.2019.8825079
20. Zhang, X., Pipattanasomporn, M., Chen, T., Rahman, S.: An IoT-based thermal model learning framework for smart buildings. IEEE Internet Things J. **7**(1), 518–527 (2020). https://doi.org/10.1109/JIOT.2019.2951106
21. Rastegarpour, S., Ghaemi, M., Ferrarini, L.: A predictive control strategy for energy management in buildings with radiant floors and thermal storage. In: 2018 SICE International Symposium on Control Systems (SICE ISCS), pp. 67–73 (2018). https://doi.org/10.23919/SICEISCS.2018.8330158
22. Qin, Y., Sheng, Q.Z., Falkner, N.J.G., Dustdar, S., Wang, H., Vasilakos, A.V.: When things matter: a survey on data-centric Internet of Things. J. Netw. Comput. Appl. **64**, 137–153 (2016). https://doi.org/10.1016/j.jnca.2015.12.016
23. Firouzi, F., Farahani, B., Marinšek, A.: The convergence and interplay of edge, fog, and cloud in the AI-driven Internet of Things (IoT). Inf. Syst. **107**, 101840 (2022). https://doi.org/10.1016/j.is.2021.101840
24. Wang, B., Liu, X., Zhang, Y.: Internet of Things. In: Wang, B., Liu, X., Zhang, Y. (eds.) Internet of Things and BDS Application, pp. 71–127. Springer, Singapore (2022). https://doi.org/10.1007/978-981-16-9194-2_2
25. Cranmer, E.E., Papalexi, M., Dieck, M.C.T., Bamford, D.: Internet of Things: aspiration, implementation and contribution. J. Bus. Res. **139**, 69–80 (2022). https://doi.org/10.1016/j.jbusres.2021.09.025
26. Rayes, A., Salam, S.: The things in IoT: sensors and actuators. In: Rayes, A., Salam, S. (eds.) Internet of Things from Hype to Reality: The Road to Digitization, pp. 63–82. Springer, Cham (2022). https://doi.org/10.1007/978-3-030-90158-5_3
27. Al-Obaidi, K.M., Hossain, M., Alduais, N.A.M., Al-Duais, H.S., Omrany, H., Ghaffarianhoseini, A.: A review of using IoT for energy efficient buildings and cities: a built environment perspective. Energies **15**, no. 16, Art. no. 16 (2022). https://doi.org/10.3390/en15165991
28. Batra, N., Singh, A., Singh, P., Dutta, H., Sarangan, V., Srivastava, M.: Data driven energy efficiency in buildings. arXiv (2014). https://doi.org/10.48550/arXiv.1404.7227
29. Yu, W., Dillon, T., Mostafa, F., Rahayu, W., Liu, Y.: A global manufacturing big data ecosystem for fault detection in predictive maintenance. IEEE Trans. Ind. Inform. **16**(1), 183–192 (2020). https://doi.org/10.1109/TII.2019.2915846

Chapter 9
Quantitative Simulation Analysis of the Function Intensity of Energy Consumption Indicators on an Office Building

Yong Ding, Weihao He, and Xue Yan

9.1 Introduction

According to the IEA's accounting results of the world's building energy consumption and CO_2 emissions in 2018, the building industry accounts for about 35% of the world's total energy consumption, of which the proportion of building operation in the total energy consumption is 30%, and the carbon dioxide emissions related to building operation in the world account for 28% of the total [1]. In the context of carbon neutralization, the requirements for improving the energy efficiency of building are growing fast at the same time. Public buildings account for a small proportion of the total building area in China, but the energy consumption per unit area is far higher than other types of buildings. As a representative of public buildings with high energy consumption, how to conduct energy management and how to reasonably evaluate the energy consumption of office buildings deserve more attention.

Through investigation, it is found that the most of the office buildings lack valid energy management. From a macro-perspective, there is no complete building energy conservation management rules and regulations system. For existing public buildings, energy conservation measures focus more on the quantity of energy consumption, yet ignoring users' demand for building environmental quality; from a micro-perspective, due to the different functionalities of public buildings and residential buildings, occupants of public buildings lack energy conservation awareness, and the implementation of energy-saving assessment mechanism and energy-saving publicity system is with limited success [2].

Hence, based on the energy management demands of office buildings, this paper takes an office building in Chongqing as an example to build energy consumption

Y. Ding (✉) · W. He · X. Yan
Chongqing University, Chongqing, China

X. Wang (ed.), *Future Energy*, Green Energy and Technology,
https://doi.org/10.1007/978-3-031-33906-6_9

model; moreover, representative energy consumption evaluation indicators are selected, based on the method of single factor sensitivity analysis, and simulation on energy consumption was carried out, as to obtain the impact of different types of indicators on energy consumption. Therefore, taking the energy-saving rate data as a reference, suggestions can be provided for judging the energy-saving potential of the office building and the direction of building energy conservation retrofitting.

9.2 Establishment of Benchmark Model

9.2.1 Selection of Energy Consumption Simulation Software

In this paper, DeST, a building energy consumption simulation software developed by the Department of Building Technology and Science of Tsinghua University, was selected for the modeling and simulation. The software organically combined and correlated the building and environmental control system with the natural room temperature as an important parameter, which could not only simulate and analyze various building performance indexes but also comprehensively analyze various passive thermal disturbances.

9.2.2 Modeling

In order to establish a typical office building model and ensure the scientific and feasibility of the research content and results, investigation and analysis of Chongqing office buildings and the study of relevant norms such as energy-saving design of public buildings were carried out, and then an office building in Chongqing was chosen as the modeling object, which is a nine-storey office building. The first floor plane shape is rectangular, with the height of 4.8 m. The second to ninth floor adopts the I-shape, and the height is 3.6 m. The offices are distributed on both sides, and the design and material characteristics of the envelope structure are set according to the survey data of the office building. The 3D model of the simulation objective is shown in Fig. 9.1.

9.2.3 Benchmark Model Parameter Settings

On the basis of determining the plane shape of the typical office model, it is necessary to assign reasonable values to the parameters of the model, such as the parameters of the envelope, indoor heat generation, and air-conditioning system.

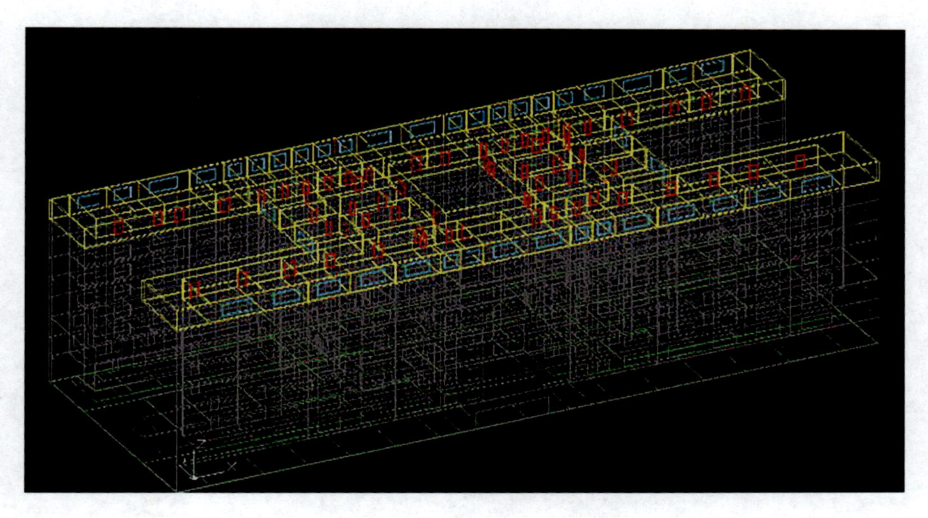

Fig. 9.1 3D model of the simulated object

Table 9.1 Geo-information parameters of simulated locations

Location	Southward angular	Latitude	Longitude	Surface reflectance
Chongqing	270.00°	29.58°	106.47°	0.30

Table 9.2 Thermal properties of external walls

Thermal-conduction resistance (m·k)/W	U-value W/(m^2·k)	Heat inertia index
0.828	1.014	3.274

This paper sets the parameters of the model in accordance with energy audit report and requirements of relevant standards.

Geographic Information Parameters Chongqing was selected as the location for this simulation, and the detailed geographic location information of the benchmark model is shown in Table 9.1.

Meteorological Parameters The simulated meteorological parameters were derived from the data provided by Chongqing Meteorological Bureau.

Setting Parameters of Building Envelope

External walls. 24 cm brick walls + polystyrene panels with internal insulation (23 mm) are selected, and its thermal properties are shown in Table 9.2.

Roof. The concrete insulation roof in the material library is selected, and the floor slab is reinforced concrete floor 150. The thermal properties of the roof are shown in Table 9.3.

Table 9.3 Thermal properties of roofs

Thermal-conduction resistance (m·k)/W	U-value W/(m^2·k)	Heat inertia index
1.522	0.595	2.837

Table 9.4 Thermal properties of windows

Glass thickness mm	Air layer mm	U-value W/(m^2·k)	SHGC	Shading coefficient	Refraction index	Emissivity
3.000	20.000	2.800	0.520	0.600	1.500	0.840

Table 9.5 Energy comsumption simulation results of an office building in Chongqing

Annual cumulative cooling load kW·h	Annual accumulated heat load kW·h	Annual maximum cooling load kW	Annual maximum heat load kW	Annual energy consumption kW·h
3750176.65	861801.97	4191.74	1574.57	530.43

Window. The double-layer aluminum alloy window is selected, and the window-wall ratio of the four orientations is set to 0.5. Relevant thermal parameters are shown in Table 9.4.

Internal disturbance. The internal disturbance parameters of each room are determined based on *Design Standards on Public Building Energy Saving (Green Building)* and software built-in recommended values [3].

Ventilation times. The ventilation times are defined as once every 2 h in the room during the day and twice every 1 h at night.

Heating and air conditioning. The office building is equipped with two sets of centrifugal chillers with the same model to meet the cooling demand. Gas boiler is used for heating. In accordance with the characteristics of this office building, the run time is set to 8:00–18:00. The set temperature of the air-conditioning system is set by the default value in terms of the room type in the DeST.

9.2.4 Simulation Results of Benchmark Energy Consumption

The total building area of the office building model is 40652.91 m^2, and the total air-conditioning area is 40403.23 m^2. The whole cooling/heating load and annual energy consumption are shown in Table 9.5.

9.3 Indicators and Analysis Method

9.3.1 Selection of Indicators

According to the categories of influencing factors, this paper divides the energy use evaluation indicators of office buildings based on energy management into two levels. The first level reflects the main classification of indicators: envelope, air-conditioning system, other equipment, and energy use behavior. The second level is the indicator layer. Based on the comprehensive consideration and feasibility of both quantitative and qualitative sides, after screening by investigation, the evaluation indicators of office building energy consumption based on the demand of energy management are determined and numbered, as shown in Table 9.6.

9.3.2 Analysis Method

Single factor sensitivity analysis can be divided into the following steps:

1. Determine analysis indicators.
2. Select the uncertain factors to be analyzed and determine the variation range of these factors.
3. Calculate the change results of the economic effect indicators of the scheme caused by the changes of various uncertain factors within the possible range of change, and establish a one-to-one corresponding quantitative relationship.
4. Determine the sensitive factors and determine the risk factors of the scheme. The methods include relative determination and absolute determination.

As regards the office building energy consumption evaluation indicators mentioned above, on the one hand, the envelope and equipment performance parameters are usually given and determined by the building inherently, which can be used to judge the building energy-saving potential and determine the direction of

Table 9.6 Indicators to evaluate office building energy consumption

One-level classification	Two-level indicators
Building envelope c_1	U-value of external walls c_{11} U-value of windows c_{12} Window-wall ratio c_{13}
Air-conditioning system c_2	COP of chillers c_{21} Set temperature of air conditioning c_{22} Set temperature of heating c_{23}
Other equipment c_3	Lighting power density c_{31} Office equipment power density c_{32}
Occupants' usage pattern c_4	Air-conditioning usage pattern c_{41} Lighting and office equipment usage pattern c_{42}

energy-saving retrofitting. On the other hand, it is an important part of building energy management to diagnose the operation status of building equipment system by combining further sub-item energy consumption information such as air-conditioning system, lighting system, etc. Moreover, the occupants' usage pattern deverses considerable attention, which can have a significant impact on the real-time operation performance.

9.4 Results and Discussion

9.4.1 Building Envelope

U-Value of External Walls Considering the existing office buildings in different construction periods and zero-energy buildings to be built in the future in Chongqing, the value range of U-value of external walls should cover all kinds of new and old specifications. Therefore, in this section, set the U-value of the exterior walls to change from 0.10 W/(m^2·K) to 1.00 W/(m^2·K), with the gradient of 0.10 W/(m^2·K), while other parameters remain unchanged. The annual building cooling and heating loads under different U-value of external walls are obtained by changing the thickness of thermal insulation materials in DeST, and then the building energy consumption is analyzed. The initial U-value of the external wall is 1 W/(m^2·K) as the benchmark. The calculation results are shown in Fig. 9.2.

With the reference of the benchmark simulation energy consumption results in Sect. 9.2.4, the corresponding energy-saving rates when the U-value of the external walls changes are obtained are as shown in Fig. 9.3.

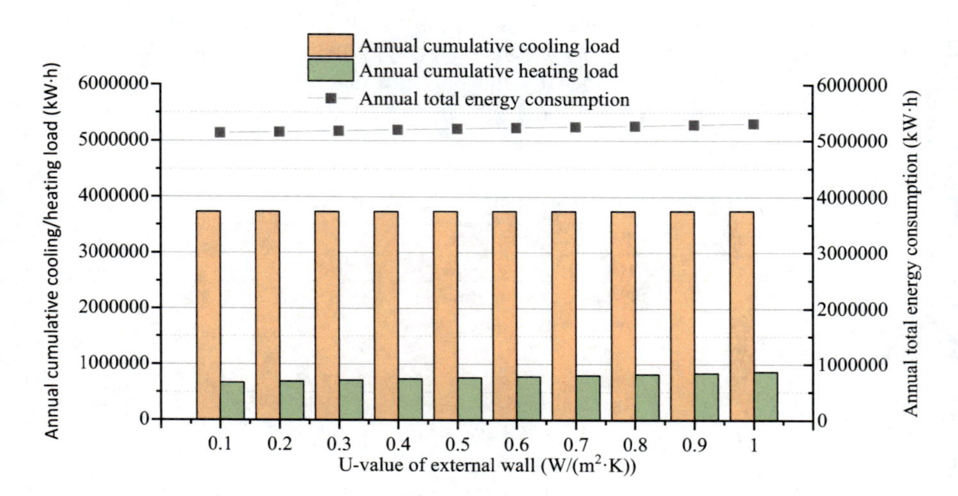

Fig. 9.2 Impact of external wall U-value on energy consumption

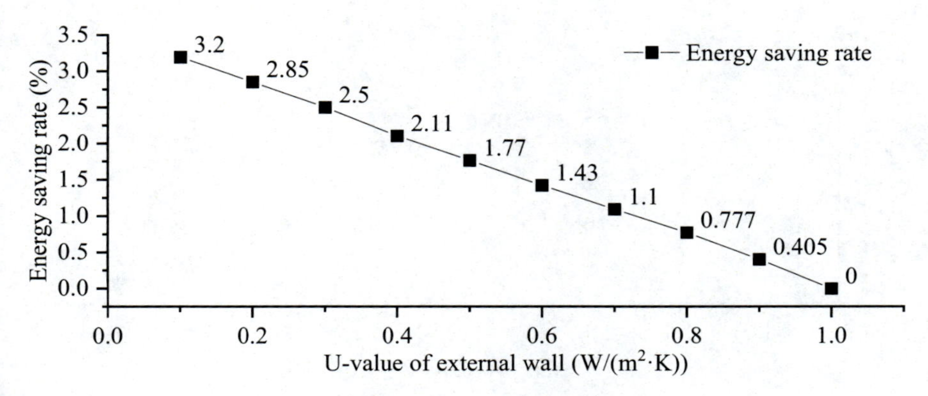

Fig. 9.3 Impact of external wall U-value on energy-saving rate

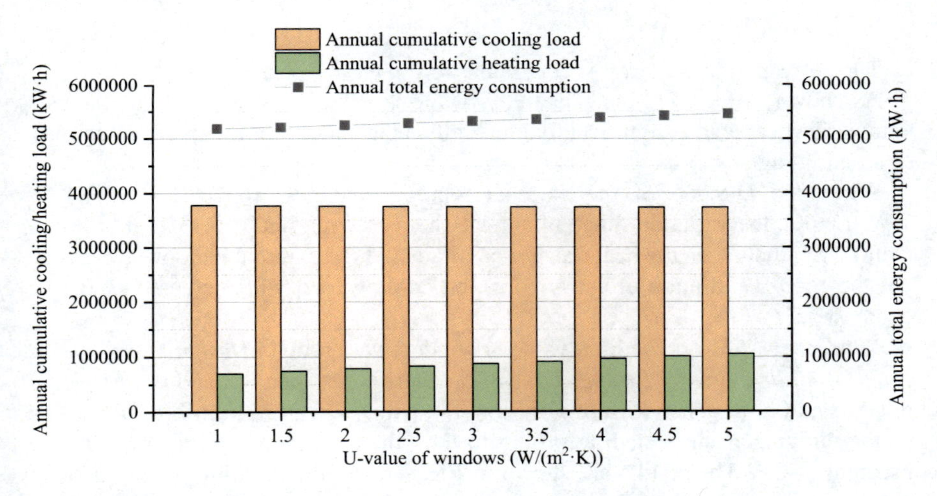

Fig. 9.4 Impact of U-value of windows on energy consumption

As regards the load, the heating load is more sensitive than cooling load to the change of the U-value of the external wall.

In terms of energy consumption, from 1.00 W/ (m^2·K) to 0.10 W/(m^2·K), the variation range of annual energy-saving rate is 3.20%.

U-Value of Windows With regard to hot summer and cold winter areas, the outer window is one of the building components with the greatest heat loss. According to *Calculation Specification for Thermal Performance of Windows, Doors and Glass Curtain-walls* [4], the range of U-value of typical windows is 0.9 W/(m^2·K) ~ 5.1 W/(m^2·K). Hence, in this section, the U-value of the outer window is changed from 1 W/(m^2·K) to 5 W/(m^2·K) in the gradient of 0.50. A total of nine different gradient parameters are set, and the initial U-value of the outer window is 2.8 W/(m^2·K) as the benchmark to analyze the impact on energy consumption. The simulation results are shown in Fig. 9.4.

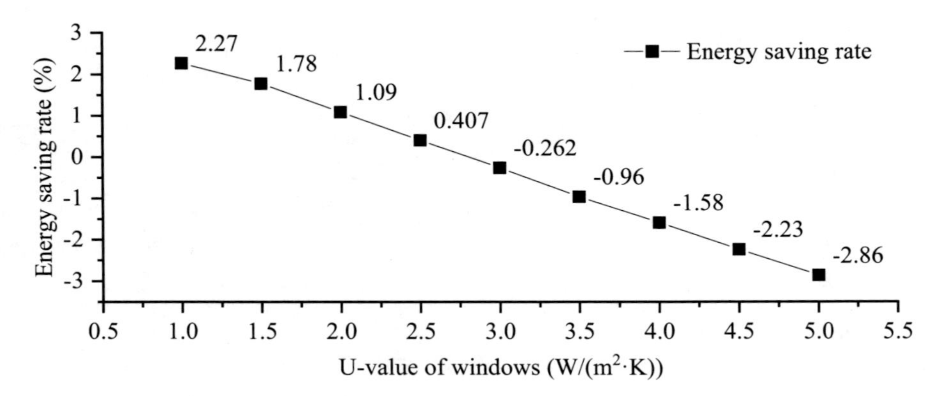

Fig. 9.5 Impact of U-value of windows on energy-saving rate

The corresponding energy-saving rates are shown in Fig. 9.5.

As shown in Fig. 9.4, the sensitivity of the U-value of windows to the annual cumulative heat load is significantly greater than the annual cumulative cooling load of the building.

When the U-value of the external window changes from 5 $W/(m^2{\cdot}K)$ to 1 $W/(m^2{\cdot}K)$, the variation range of annual energy-saving rate is 5.13%. It is worth mentioning that when the heat transfer coefficient of the external window decreases, yet the room ventilation effect is poor, the cooling load will increase instead of decrease [5].

Window-wall ratio. The limit value of window-wall ratio in *Design Standards on Public Building Energy Saving (Green Building)* is between 0.2 and 0.8 and shall not exceed 0.7 in general. In this section, the window-wall ratio in four directions of the building is simulated according to the scheme of from 0.2 to 0.7 with the gradient of 0.1. The results are shown in Fig. 9.6. As the building model plane is I-shaped, and the building faces south, the area of eastward and westward windows is much smaller than that of southward and northward windows, resulting in the change of wall ratio from west and east has no obvious impact on the building energy consumption.

Compared to the window-wall ratio benchmark value of 0.5, the maximum energy-saving rates in the east, south, west, and north are 0.88%, 2.40%, 1.10%, and 0.80% respectively, as shown in Fig. 9.7. The sensitivity of the four orientation windows and walls to the annual energy consumption of the building is ranked as follows: southward>westward>eastward>northward.

9.4.2 Air-Conditioning System

Chiller COP The chiller system is mainly composed of chiller units, pumps, and cooling towers, whose energy consumption accounts for 62%, 30%, and 8%,

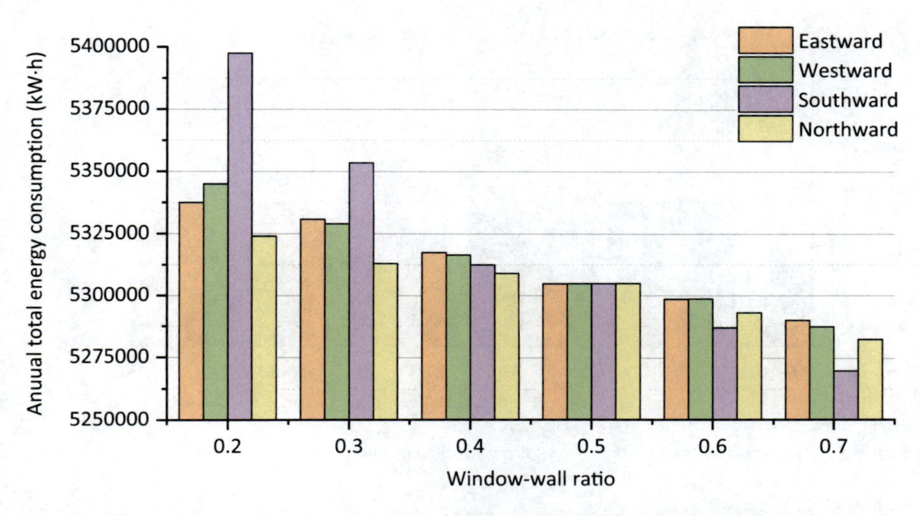

Fig. 9.6 Simulation results of building annual energy consumption under different window-wall ratio

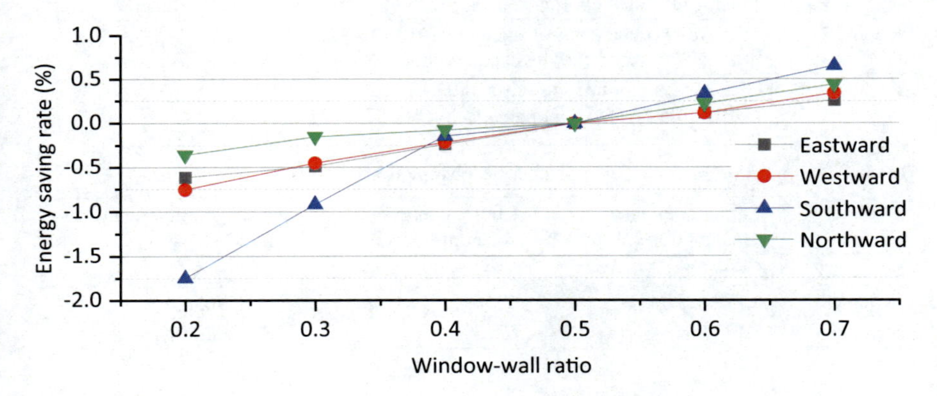

Fig. 9.7 Impact of window-wall on energy-saving rate in different orientations

respectively [6]. Therefore, reducing the energy consumption level of the unit is the key to the energy management of the air-conditioning system. The total cooling capacity of the screw chiller used in the building is 3866 kW; according to simulation, the time frequency statistics of partial load rate of the system are shown in Fig. 9.8.

The system load rate is an important factor influencing the COP of chiller. In order to keep the chiller units running at a high COP, as to see the impact of different chiller schemes and actual COP on energy consumption, three kinds of chiller schemes are proposed according to the peak cooling load of the building. The chiller sample is chosen from built-in database in DeST, and all the chiller schemes are shown in Table 9.7.

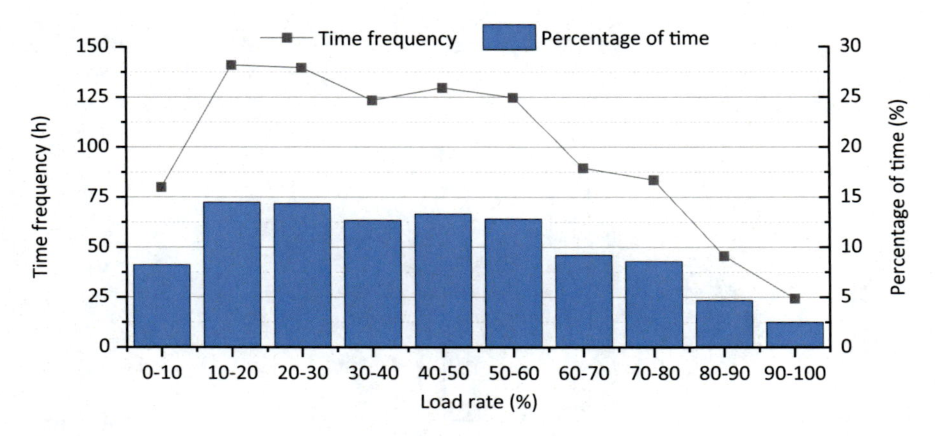

Fig. 9.8 Time frequency statistics of system partial load rate

Table 9.7 Chiller schemes

No. of sheme	Parameters of chillers	Remarks
Scheme 1	TFLX-500 (rated cooling capacity: 1933 kW)*2	Benchmark scheme
Scheme 2	TFLX-600 (rated cooling capacity: 1933 kW)*2; LSBLX826 (rated cooling capacity: 868 kW)*1	
Scheme 3	CHB3 (rated cooling capacity: 1407 kW)*2; LSBLX1052 (rated cooling capacity: 1052 kW)*1	

Table 9.8 Impact of chiller scheme on energy consumption

Chiller scheme	Annual total energy consumption kW·h	Chiller energy consumption kW·h	Refrigeration quantity kW·h
Scheme 1	5,304,303	720385.62	3274615.63
Scheme 2	5,178,421	635697.73	3273965.70
Scheme 3	5,170,906	622376.65	3272023.91

As shown in Table 9.8, the simulation results show that Scheme 2 and Scheme 3 have better performance on energy conservation, as the annual energy-saving rate is 2.37% and 2.51%, respectively, compared with Scheme 1; moreover, annual energy-saving rates of the chillers in Scheme 2 and Scheme 3 are 11.76% and 13.60%, respectively.

According to simulation, in August 23, the cooling load of the day is between 1000 kW and 2000 kW. As shown in Fig. 9.9, when the load increases to 1721 kW at 15:00, the COP of chillers in Scheme 3 reaches the highest, which is superior to Scheme 2 and Scheme 1. Therefore, in terms of deploying chillers, it is necessary to consider selecting the corresponding refrigeration scheme according to the proportion of the duration of partial load.

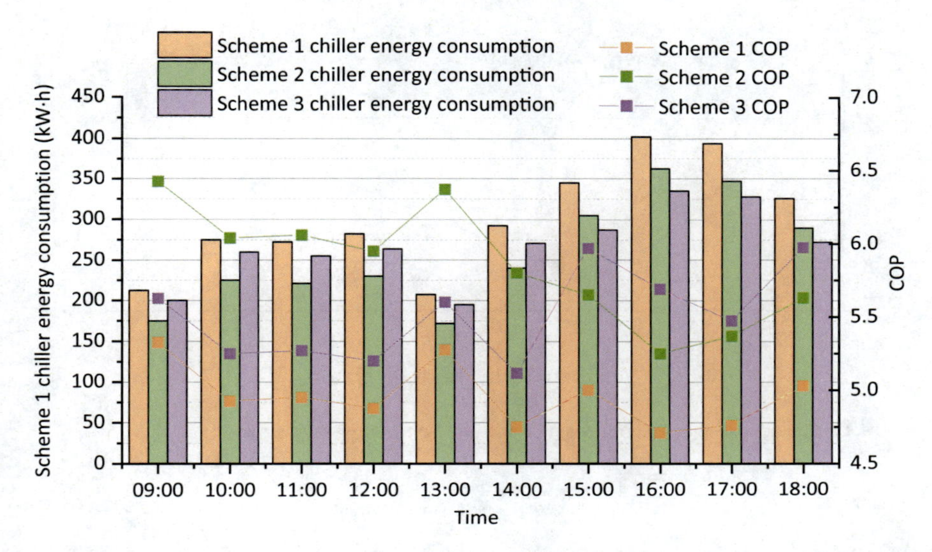

Fig. 9.9 Comparison of the working performance of the chillers from 9 am to 6 pm on August 23 from the three schemes

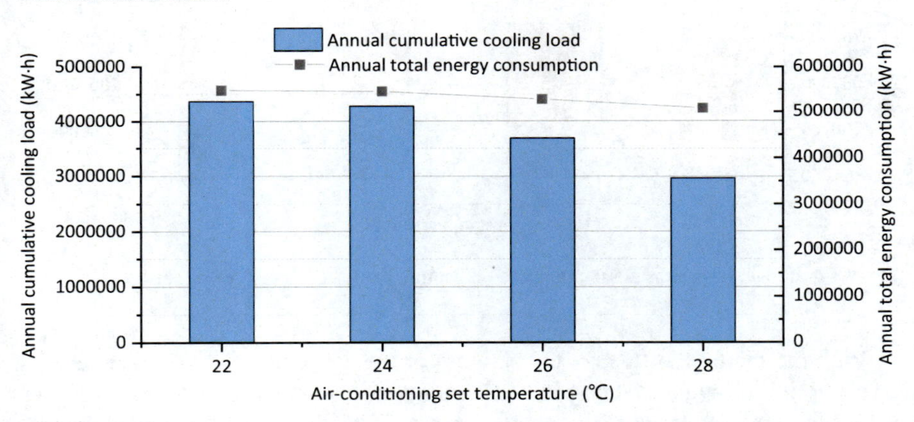

Fig. 9.10 Impact of air-conditioning set temperature on cooling load and energy consumption

Air-Conditioning/Heating Set Temperature In the light of the provisions of *Design Code for Heating Ventilation and Air Conditioning of civil buildings* on indoor design working conditions [7], the reference temperature range for thermal comfort in Chongqing is 22 °C–28 °C in summer and 18 °C–24 °C in winter. Hence, in this section, as for indoor set temperature, 22 °C, 24 °C, 26 °C, and 28 °C are set in summer, and 18 °C, 20 °C, 22 °C, and 24 °C are set in winter, so as to take four different working conditions that are set in winter and summer, respectively, to study the effect of temperature setting on energy consumption.

Figures 9.10 and 9.11 show the simulation results of energy consumption and energy-saving rate when the set temperature of air conditioner changes.

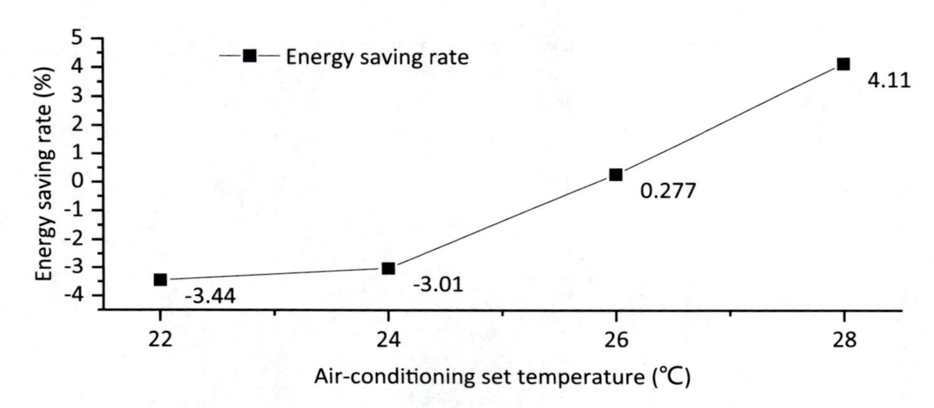

Fig. 9.11 Impact of air-conditioning set temperature on energy-saving rate

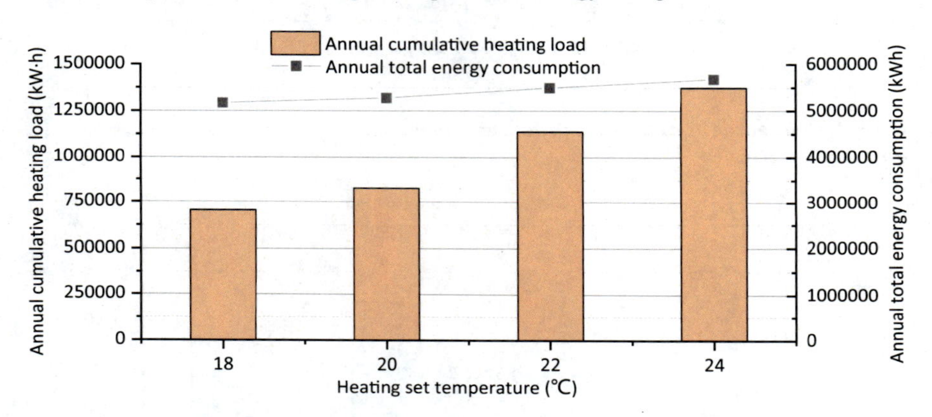

Fig. 9.12 Impact of heating set temperature on heating load and energy consumption

With the increase of the set temperature of the air conditioner, the total energy consumption throughout the year takes on a decreasing change. As shown in Fig. 9.11, when the set temperature changes from 22 °C to 28 °C, the annual energy-saving rate increased from −3.44% to 4.11%.

The energy consumption and energy-saving rate simulation results of heating set temperature change are shown in Figs. 9.12 and 9.13.

With the increase of the heating set temperature, the annual cumulative heat load and the annual total energy consumption increase gradually. As shown in Fig. 9.13, when the heating set temperature changes from 18 °C to 24 °C, the annual energy-saving rate was reduced from 2.53% to −7.11%.

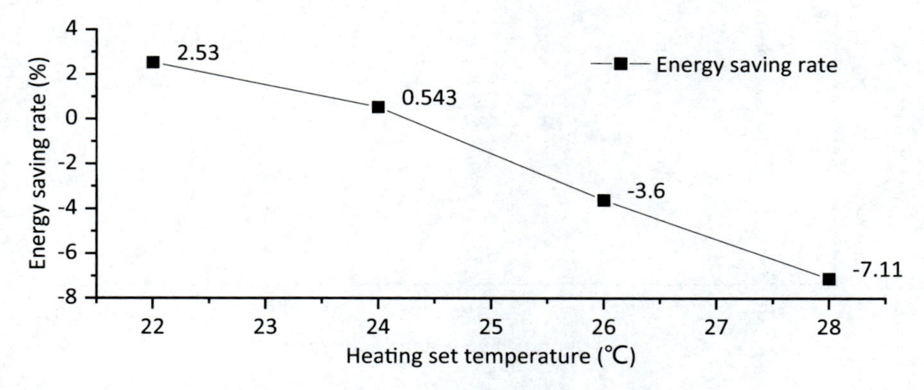

Fig. 9.13 Impact of heating set temperature on energy-saving rate

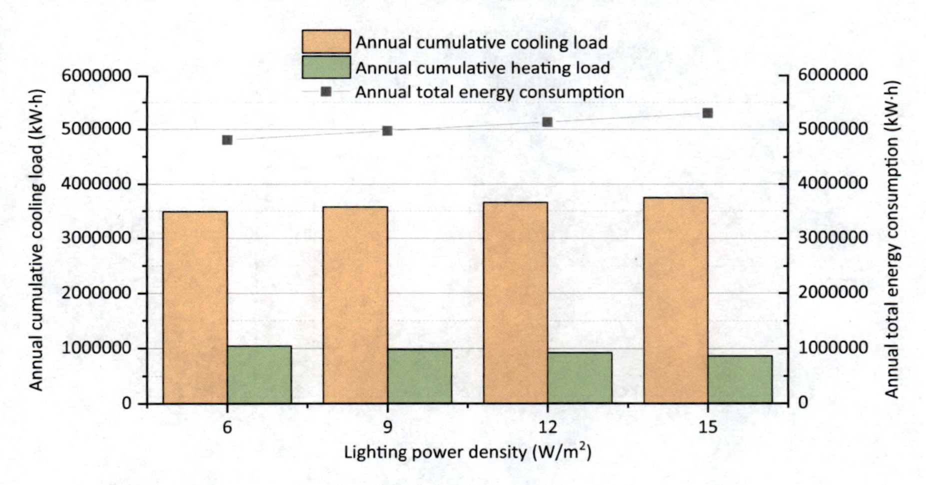

Fig. 9.14 Impact of lighting power density on energy consumption

9.4.3 Other Equipment

Lighting Power Density In office buildings, lighting power density is often used to reflect the energy consumption of the lighting system. Considering that some existing office buildings in Chongqing were built a decade ago and earlier, so as for simulation, lighting power density is set according to the limit of 15 W/m^2 in accordance with the standard GB 50034–2013. In this section, four different lighting power density levels with a gradient of 3 W/m^2 from 6 W/m^2 to 15 W/m^2 are set for offices and meeting rooms. The simulation results are shown in Figs. 9.14 and 9.15.

As shown in Figs. 9.14 and 9.15, when the lighting power density changes from 6 W/m^2 to 15 W/m^2, the annual energy consumption increases by 9.44%.

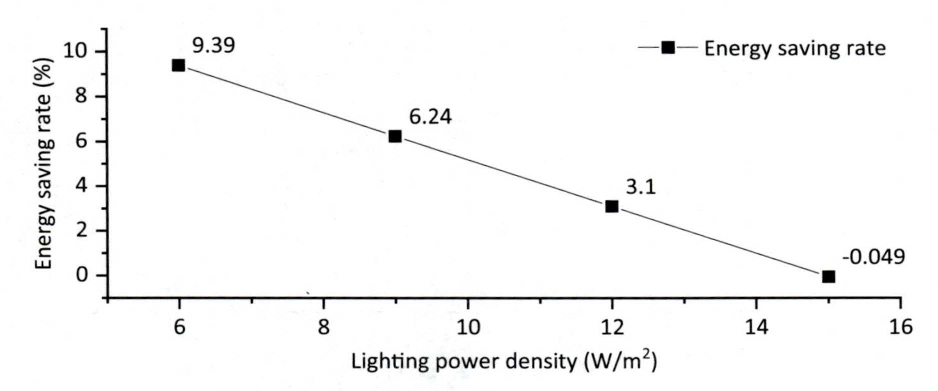

Fig. 9.15 Impact of lighting power density on energy-saving rate

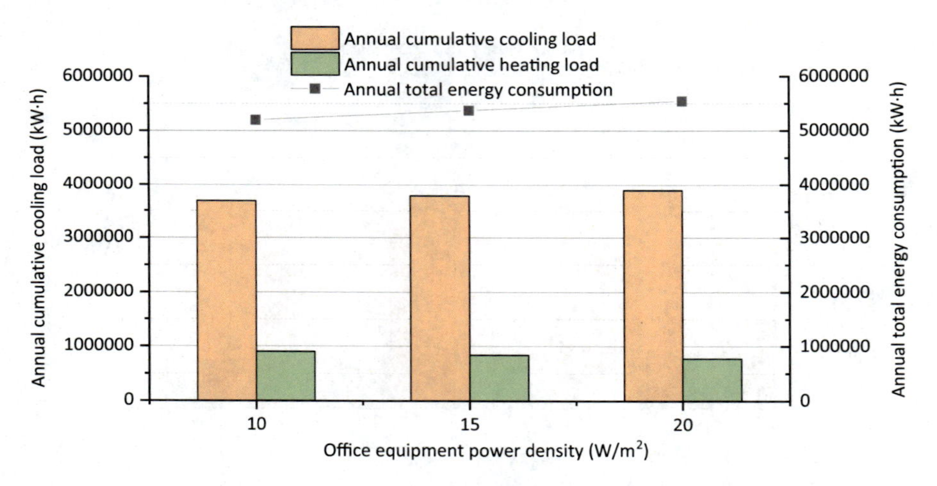

Fig. 9.16 Impact of office equipment power density on energy consumption

Office Equipment Power Density According to older provision DBJ50–052-2013, the equipment power density shall be less than 20 W/m^2. Therefore, in this section, three different office equipment power densities of 10 W/m^2, 15 W/m^2, and 20 W/m^2 with a gradient of 5 W/m^2 are set to analyze the impact of office equipment power density on energy consumption. The simulation results are shown in Figs. 9.16 and 9.17.

Comparing the simulated annual energy consumption results with the initial parameters, it can be seen that when the equipment power density changes from 10 W/m^2 to 20 W/m^2, the annual energy consumption increases by 6.52%. Additionally, the influence of equipment power density on energy consumption presents certain linearities. Therefore, in the premise of meeting functionality, the office equipment power density should be limited as much as possible.

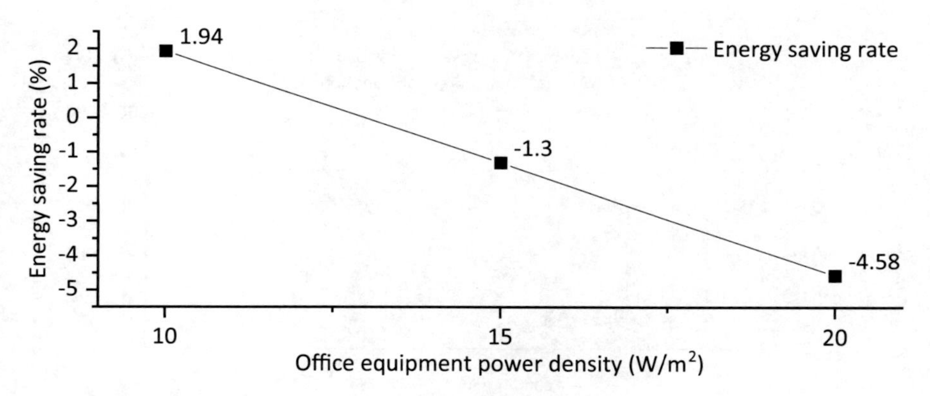

Fig. 9.17 Impact of office equipment power density on energy-saving rate

9.4.4 Occupants' Usage Pattern

Air-Conditioning Usage Pattern Through the investigation of office buildings in Chongqing, it is found that there are three typical usage patterns of air conditioning: Occupants turn on the air conditioner when feeling hot and turn it off when feeling cold; occupants turn on the air conditioner when entering the office and turn it off when leaving the office; and air conditioner is always being turned on during daytime.

Research shows that the acceptable indoor air temperature distribution of 80% of people is 24.2 °C ~ 29.8 °C, and the thermal neutral temperature of residents in hot summer and cold winter areas is 27.3 °C in summer [8]. The indoor heating temperature in Sichuan Basin and Chongqing is 15.3 ~ 16.0 °C in winter, and the lower limit of temperature is 15.3 °C as the indoor design temperature. According to the above analysis, in order to analyze the impact of air-conditioning usage pattern on energy consumption, this paper summarizes the air-conditioning usage pattern into two extreme modes.

Mode 1: When the indoor temperature reaches 27.3 °C in summer, turn on the air-conditioner, and when the indoor temperature is lower than 15.3 °C in winter, heating will be started.

Mode 2: The air conditioner operates from 8:00 to 17:00 with no downtime, and the set temperature is the recommended temperature for large office buildings, i.e., 26.0 °C in summer and 20.0 °C in winter.

In DeST, the annual cumulative cooling and heating load and the annual total energy consumption results obtained by setting the room's air-conditioning schedule are shown in Figs. 9.18 and 9.19.

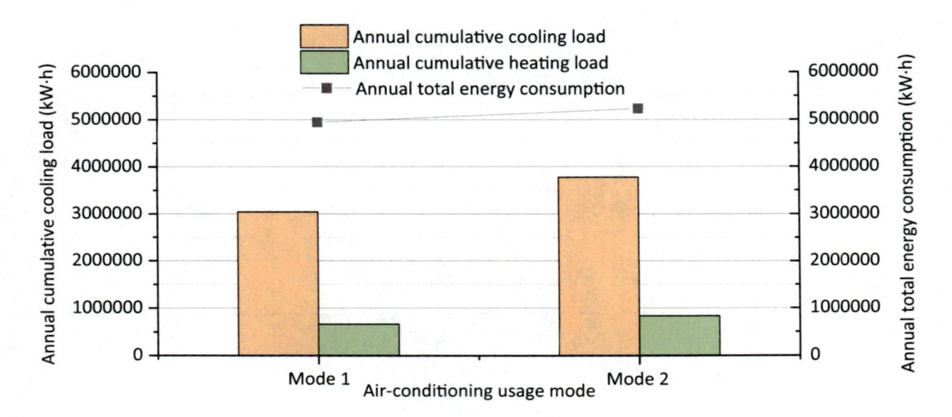

Fig. 9.18 Impact of air-conditioning usage mode on energy consumption

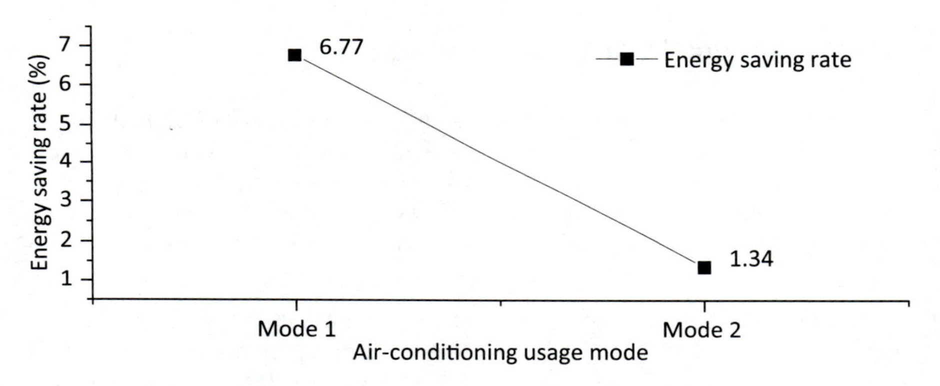

Fig. 9.19 Impact of air-conditioning usage mode on energy-saving rate

When the air-conditioning usage mode switches from mode 1 to mode 2, it can be found that compared with mode 1, the total annual energy consumption of mode 2 increased by 5.82%.

Based on the simulation results of the benchmark model, it can be seen that the annual energy-saving rates of the two modes are 6.77% and 1.34%, respectively. The usage pattern of air conditioning has a great impact on the annual energy consumption; hence it is necessary to continue to publicize the awareness of keeping the air conditioner being switch off during nonworking hours for occupants.

Lighting and Office Equipment Usage Pattern From the perspective of light and equipment's on-time, there are two typical modes of using lights and office equipment: (1) economical, wherein occupants turn on the lights and office equipment only between entering and leaving the working zone, and (2) improvident, wherein occupants always have lights and office equipment on in working time. Furthermore, the lighting and office equipment usage pattern can be summarized into two different modes.

Mode 1: Control the opening ratio of lights/equipment in line to the actual occupants' presence. According to the utilization rate model of typical office occupants [9], the average indoor stay time is 6 h. So for this mode, 8:00 ~ 11:00 is set as the working time, and all occupants are in the room. 11: 00 ~ 14:00 is set for dining and resting. At this time period, the utilization rates of the lamps/equipment are set to 0.3. During the time period from 14:00 to 16:00, the utilization rates of the lights/equipment are set as 1, and the utilization rates are set as 0.5 from 16:00 to 18:00 in consideration of occupants' turnover and overtime.

Mode 2: Set all lights and office equipment to be turned on in the working hours from 8:00 to 18:00.

After setting all the parameters in DeST, the simulation results are shown in Figs. 9.20 and 9.21.

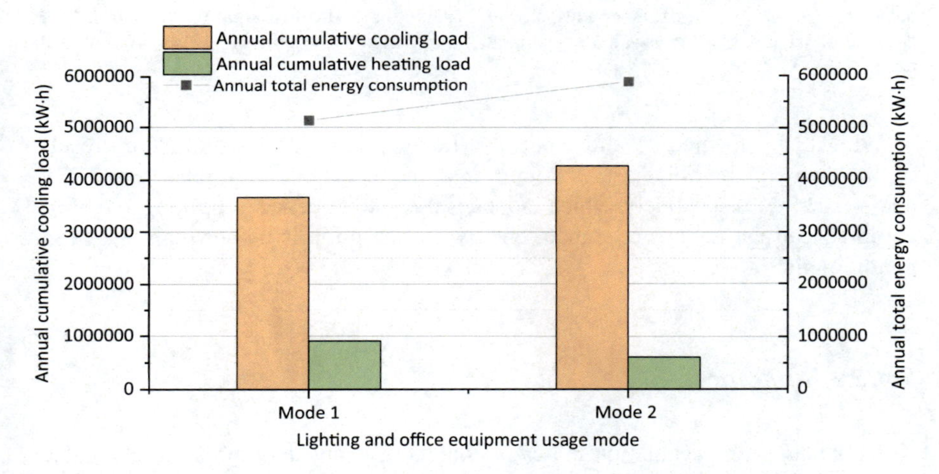

Fig. 9.20 Impact of lighting and office equipment usage mode on energy consumption

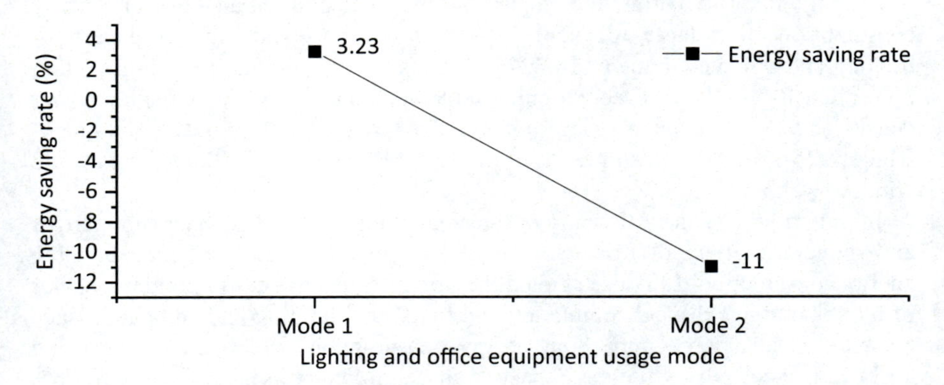

Fig. 9.21 Impact of lighting and office equipment usage mode on energy-saving rate

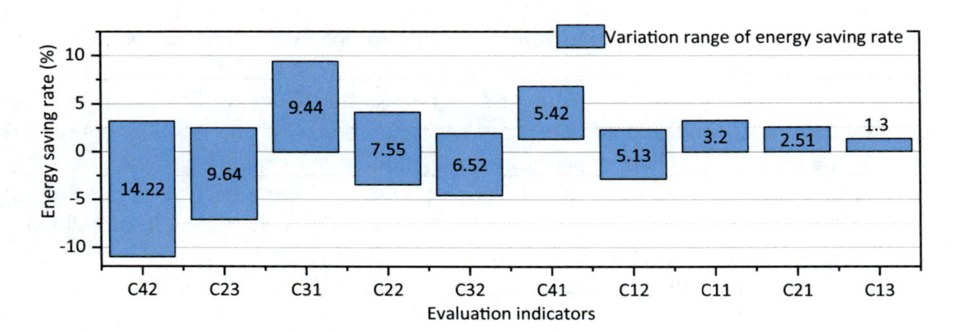

Fig. 9.22 Variation range of energy-saving rate of evaluation indicators. C42: occupants' usage pattern of lighting and office equipment; C23: set temperature of heating; C31: lighting power density; C22: set temperature of air conditioning; C32: office equipment power density; C41: occupants' usage pattern of air conditioning; C12: lighting and office equipment control; C12: U-value of windows; C11: U-value of external walls; C21: COP of chillers; C13: Window-wall ratio

With regard to mode 1 and mode 2, when compared to the simulation results of the benchmark value, it is found that the energy-saving rates are 3.23% and − 10.99%, respectively, which indicates that the influence of lighting and office equipment usage pattern on annual energy consumption is higher than that of air conditioning.

9.4.5 Summary

The impacts of the evaluation indicators on energy consumption have been calculated and analyzed. Now it is necessary to make a summary, as shown in Fig. 9.22.

Among this simulation, the impact of the evaluation indicators on energy consumption from large to small is as follows: occupants' usage pattern of lighting and office equipment (14.22%), heating set temperature (9.64%), lighting power density (9.44%), air-conditioning set temperature (7.55%), equipment power density (6.52%), occupants' usage pattern of air conditioning (5.42%), U-value of windows (5.13%), U-value of external walls (3.20%), COP of chillers (2.51%), and window-wall ratio (1.30%).

In order to verify the rationality of the simulation results, this paper summarizes and refines a variety of researches on the influential factors of office building energy consumption. Ling has studied the impact of energy consumption behavior of office buildings through simulation and found that lighting and computer usage behavior have a large impact on energy consumption, which are 16.79% and 14.21%, respectively, while the impact of air-conditioning usage pattern is 13.06% [10]. Zhang compared the energy consumption of office buildings with and without lighting control measures through energy consumption simulation and found that the energy-saving rate of lighting system energy-saving control is between 13.0% and 14.6% [11]. Liu simulated the energy consumption of office buildings by designing orthogonal tests, and the results show that lighting power density and air-conditioning set temperature have a greater impact on energy consumption, more than COP of chiller, external walls heat transfer coefficient, and window-wall ratio [12]. In a few words, the fact that lighting usage pattern has a significant impact on building energy consumption is with reliability and credibility.

9.5 Conclusion

Finally, it can be concluded as below:

With regard to office building's energy management, compared with the improvement of common quantifiable factors (e.g., building envelope and equipment parameters), good user behaviors make appreciable energy-saving effect and worth generalization; that is to say, although good commissioning can ensure that the performance of main energy consumption equipment and systems reaches a good state, occupants as subjective individuals have great potential in achieving energy conservation. All in all, it counts for much to cultivate users' awareness of energy conservation and establish an incentive system.

However, the method used in this paper still remains to be improved in some certain ways. For instance, the size, structure, and orientation of the building model would have influence on the variation range of energy-saving rate of the U-value of external window, window-wall ratio, user behavior on lighting, and other indicators. Moreover, in this very case, the range of lighting and equipment power density is determined by relevant provisions and researches, while deviation may occur in the actual projects. In brief, with the support of a large number of real sampled data, the accuracy of the calculation results can be improved certainly.

References

1. Building Energy Efficiency Research Center of Tsinghua University: Research Report on the Annual Development of Building Energy Efficiency in China, 1st edn. China Building Industry Press, Beijing (2020)
2. Cheng, Z., Huafang, G.: Current situation and prospect of energy management in public buildings. Guangxi Electric Power. **156**(4), 35–37 (2017)
3. Dong, Y., Dong, M., Xie, Z.: Design standards on public building energy saving(Green building), 2nd edn. China Architecture & Building Press, China (2020)
4. Yang, S., Lin, H.: Qinglin Meng: Calculation specification for thermal performance of windows, doors and glass curtain-walls, 1st edn. China Architecture & Building Press, China (2008)
5. Huang, Y.J., Mankibi, M.E., Cantin, R.: Application of fluids and promising materials as advanced inter-pane media in multi-glazing windows for thermal and energy performance improvement: a review. Energ. Buildings. **253**(12), 1–14 (2021)

6. Huang, J.'e., Feng, W.: A calculation model for energy consumption of screw refrigeration unit. HVAC. **37**(4), 121–126 (2007)
7. Wei, X., Zou, Y., Hongqing, X.: Design code for heating ventilation and air conditioning of civil buildings, 2nd edn. China Architecture & Building Press, China (2012)
8. Li, J., Yang, L., Liu, J.: Investigation and study on indoor thermal comfort of summer residence in hot summer and cold winter areas. Sichuan Architec. Sci. Res. **34**(4), 200–205 (2008)
9. Luo, T., Yan, D., Jiang, Y., Zhao, J.: Study on simulation method of lighting energy consumption in office buildings. Build. Sci. **33**(04), 101–109 (2017)
10. Ling, W., Cui, Q., Song, B., Deng, Q., Zhu, X.: Exploring ways of building energy conservation driven by user behavior. Build. Sci. **34**(12), 130–139 (2018)
11. Zhang, Q., Lin, B.: Verification and sensitivity analysis of energy consumption difference of typical green office buildings. HVAC. **49**(08), 31–39 (2019)
12. Liu, D., Chen, Q., Sen, Y., Muda, Y.: Building energy consumption analysis and modeling based on data. J. Tongji Univ. Nat. Sci. Ed. **38**(12), 1841–1845 (2010)

Chapter 10
Preliminary Multiple Linear Regression Model to Predict Hourly Electricity Consumption of School Buildings

Keovathana Run, Franck Cévaër, and Jean-François Dubé

10.1 Introduction

Approximately 40% of EU energy consumption and 46% of energy-related greenhouse gas emissions are attributable to buildings. Nearly 75% of the building stock in the EU is now energy inefficient, and about 35% of structures are older than 50 years [1]. According to data released by the Agency for the Environment and Energy Management (ADEME), the building sector in France is responsible for 25% of CO_2 emissions and 44% of energy usage.

Researchers discovered that energy forecasting techniques that use historically recorded time series energy data have enormous value in energy optimization for existing buildings [2]. Data-driven models have gained popularity among academics due to their simplicity, ability to handle large data sets, and high prediction accuracy, though this is not true for all types of data-driven models [3].

The objective of this paper is to develop a preliminary Multiple Linear Regression (MLR) model that aims to predict the electric power consumption per hour on school buildings. Two initial models were compared to see their prediction ability, namely, the first-order model and the two-way interaction model using the forward regression that is described in Sect. 10.2.2 and the model evaluation metrics in Sect. 10.2.3.

K. Run (✉) · F. Cévaër · J.-F. Dubé
LMGC, University of Montpellier, Montpellier, France
e-mail: keovathana.run@umontpellier.fr

X. Wang (ed.), *Future Energy*, Green Energy and Technology,
https://doi.org/10.1007/978-3-031-33906-6_10

10.2 Method

10.2.1 MLR

The method in this study is based on a multivariate regression analysis, which accounts for the variation of the independent variables in the dependent variables synchronically [4]. The multiple linear regression model is:

$$Y_i = \beta_0 + x_{i,1}\beta_1 + x_{i,2}\beta_2 + \cdots + x_{i,p}\beta_p + e_i$$

where

Y is the response (dependent variable)
x is the predictors (independent variables)
β is the unknown regression coefficients
e is unknown errors

$$i = 1, \ldots, n$$

n is the sample size
e_i is the error to account for the discrepancy between predicted and the observed data.

After the models are developed and checked, the predicting is then made. All too often the MLR model seems to fit the "training data" well, but when new "testing data" is collected, a very different MLR model is needed to fit the new data well. Therefore, it is important to wait until after the MLR model has been showed to make good predictions before claiming that the model gives good predictions [5].

10.2.2 Predictor Selection

The potential predictors are preselected. This study employs forward regression, which starts with a model with no predictors, to choose its predictors (the intercept only model). After that, variables are added to the model one at a time until none more can improve it by a particular standard. The variable that significantly improves the model is introduced at each step. A variable stays in the model once it is added.

In a model, to predict a response vector Y in $\mathbb{R}^n$ (preserve original elements), the predictor matrix X should be in $\mathbb{R}^{np}$ (preserve original elements), and yields a subset of each size $k = 0, \ldots, \min\{n, p\}$ (preserve original elements). Formally, the procedure starts with an empty active set $A = \{0\}$, and for $k = 0, \ldots, \min\{n, p\}$, selects the variable indexed by equation (10.1) that leads to the lowest squared error when added to A_{k-1}, or equivalently, such that X_{j_k}, achieves the maximum absolute correlation with Y, after we project out the contributions from $X_{A_{k-1}}$. A note on notation: here we write $X_S \in \mathbb{R}^{n\,|\,S|}$ for the submatrix of X whose columns are indexed by a set S (and when $S = \{\, j \,\}$, we simply use X_j).We also write P_s for the projection matrix onto the column span of X_s and $PS^{\perp} = IP_s$ for the projection matrix onto the orthocomplement. At the end of step k of the procedure, the active set is updated, $Y_k = A_k - 1 \cup \{j_k\}$, and the forward stepwise estimator of the regression coefficients is defined by the least squares fit onto X_{A_k} [5].

$$j_k = {}_{j \notin A_{k-1}}{}^{\text{argmin}} \parallel Y - P_{A_{k-1} \bigcup \{j_k\}} Y \parallel_2^2 = {}_{j \notin A_{k-1}}{}^{\text{argmin}} \frac{X_j^T P_{A_{k-1}}^{\perp} Y}{\parallel P_{A_{k-1}}^{\perp} X_j \parallel_2} \tag{10.1}$$

10.2.3 Model Selection

The performance of the models is assessed using tenfold cross-validation after the best models from the combination of chosen parameters have been developed. Regression analysis uses the mean absolute error (MAE), mean squared error (MSE), root mean squared error (RMSE), and mean absolute percentage error (MAPE) metrics to measure each model's prediction error. The better the model, the lower the MSE, MAE, and RMSE. Coefficient of determination (R^2) denotes the relationship between the values of the desired outcomes and those that the model predicts. The better the model, the greater the R^2.

– MAE represents the difference between the original and predicted values extracted by averaged the absolute difference over the data set.

$$\text{MAE} = \frac{1}{N} \sum_{i=1}^{N} |y_i - \hat{y}_i|$$

– MSE is the average of the squared difference between the original and predicted values in the data set. It measures the variance of the residuals.

$$\text{MSE} = \frac{1}{N} \sum_{i=1}^{N} (y_i - \hat{y}_i)^2$$

– RMSE is the error rate by the square root of MSE. It measures the standard deviation of residuals.

$$\text{RMSE} = \sqrt{\frac{1}{N} \sum_{i=1}^{N} (y_i - \hat{y})^2}$$

– MAPE is the percentage error calculated in terms of absolute errors, without regard to sign.

$$\text{MAPE} = \frac{100}{N}\sum_{i=1}^{N} \mid \frac{y_i - \hat{y}_i}{y_i} \mid$$

- R^2 represents the coefficient of how well the values fit compared to the original values. The value from 0 to 1 is interpreted as percentages.

$$R^2 = \frac{\sum (y_i - \hat{y}_i)^2}{\sum (y_i - \overline{y_i})^2}$$

where $\hat{y}$ and $\overline{y}$ are, respectively, predicted and mean value of y measured value at the ith moment, and N represents the number of predictions.

10.3 Case Study: IUT de Nîmes School Buildings

As part of the OEHM project, IUT de Nîmes campus was selected as a case study for the Ph.D. thesis of the first author. This campus is in the south of France, at 43°49′N longitude and 4°19′E latitude. The climate of this region is classified as CSA, with relatively mild winters and hot summers, often referred to as " Mediterranean " according to Koeppen and Geiger [6]. Since 2019, 338 sensors in total of 6 types (Elsys, Class'Air, CM868LR, IR868LR, BT1-L, and Adeunis) have been placed on the site.

The data from three buildings built in 1969 were collected. They are civil engineering building (GC), electrical engineering building (GEII), and material engineering building (GMP) with total net surface area of 4762 m^2, 3627 m^2, and 6357 m^2, respectively. Naturally ventilated, they also have the same floor plans, a two-story teaching building and a one-story workshop building with a high ceiling. The heating system is hot water radiators, supplied with heat by the urban heating network. The electrical energy is dedicated to the rest of the appliances in the buildings including lighting, the electrical distributor, electrical radiators, air conditioners (reversible), etc. Therefore, the electricity consumption still depends on weather conditions and the indoor climate.

The sensors record and transmit every 15 min for indoor carbon dioxide (CO_2), indoor temperature (T_{in}), and indoor relative humidity (HR_{in}) and every 1 h for real-time electricity consumption. The outdoor temperature (T_{ext}), outdoor relative humidity (HR_{ext}), and global solar radiation (SR) are taken every 1 h from the nearest representative station, the climate data of Nîmes Courbessac from Météo France. The analysis is done during 5 months, from November 2021 to April 2022, when all necessary data were available. Each parameter's time basis was reset to every hour using time interpolation.

Table 10.1 shows the range and variation of each parameter of each building. It is evident that the outdoor weather is between −3.5 °C and 27.5 °C and highest

Table 10.1 Statistics of collected data set

Parameters	Building	Min	Median	Mean	Max	SD[a]	SE[b]
Consumption (kWh/h)	GC	5.00	10.00	15.99	80.00	12.06	0.20
	GEII	0.00	5.00	10.06	81.00	10.82	0.18
	GMP	9.00	18.00	25.43	93.00	16.02	0.26
CO_2 (ppm)	GC	378.79	440.27	472.04	1084.26	90.63	1.49
	GEII	368.51	437.32	470.37	1042.86	92.74	1.53
	GMP	373.21	457.32	551.41	1727.71	199.10	3.28
T_{in} (°C)	GC	14.55	19.60	19.47	22.84	1.32	0.02
	GEII	13.47	20.67	19.81	23.33	2.49	0.04
	GMP	17.38	21.74	21.48	24.12	1.09	0.02
HR_{in} (%)	GC	22.53	37.00	37.70	55.67	7.43	0.12
	GEII	18.00	36.00	37.61	69.00	10.96	0.18
	GMP	23.92	36.05	37.25	54.90	6.84	0.11
T_{ext} (°C)		−3.50	8.70	9.04	27.50	4.85	0.08
HR_{ext} (%)		18.00	65.00	64.90	97.00	18.67	0.31
SR (MJ/m^2)		0.00	0.00	0.40	3.28	0.67	0.01

[a]*SD* standard deviation, [b]*SE* standard error

temperature indoor is between 17.38 °C and 24.12 °C. Peak value for global solar radiation is 3.28 MJ/m^2. From a quick analysis, GMP has the most corresponding variations for all the parameters.

10.4 Model Development

10.4.1 Preselection Variables

The preselected explanatory and dependent variables for the models are as follows:

(i) Dependent variable: Y = Hourly electricity usage (kWh/h)
(ii) Predictor variable 1: x1 = CO_2 (ppm)
(iii) Predictor variable 2: x2 = T_{in} (°C)
(iv) Predictor variable 3: x3 = HR_{in} (%)
(v) Predictor variable 4: x4 = T_{ext} (°C)
(vi) Predictor variable 5: x5 = HR_{ext} (%)
(vii) Predictor variable 6: x6 = SR (MJ/m^2)

To get more reliable results, another three proxy variables are added: day index, hour index, and building net floor area.

(i) Predictor variable 7: x7 = Day Index (Weekday/Weekend)

(ii) Predictor variable 8: x8 = Hour Index (daytime 7 h00 – 19 h00/ nighttime 19 h00 – 6 h00)
(iii) Predictor variable 9: x9 = Building net floor area (m^2)

10.4.2 Model Selection

The initial model using first order of all nine predictors can be expressed as:

$$Y_1 = \beta_0 + x_1\beta_1 + x_2\beta_2 + x_3\beta_3 + x_4\beta_4 + x_5\beta_5 + x_6\beta_6 + x_7\beta_7 + x_8\beta_8 + x_9\beta_9$$

The analysis is done using Rstudio version 4.0.3. (2020-10-10). The regsubsets() function [leaps package] computes the forward regression; the tuning parameter nvmax specifies the maximum number of predictors to incorporate in the model. It returns a variety of models in sizes ranging from small to large. The performance of the models is then carefully compared to select the best one. Among the nine models returned, the best performance is when all nine variables are considered, with RMSE of 0.105 and R^2 of 0.60. However, the value of R^2 of the trained model is rather weak, meaning that it can only explain 60% of the variance. Therefore, a two-way interaction of this trained model is carried out for an equation that can be written as:

$$Y_2 = \beta_0 + x_1\beta_1 + \cdots + x_9\beta_9 + x_1x_2\beta_{10} + \cdots + x_8x_9\beta_{45}$$

where $i = 1, \ldots, 8$ and β_0 is the intercept value. Using the same forward regression to find the best combination of predictors, 45 models are returned and the best model of 40 variables has RMSE of 0.08 and R^2 of 0.73. The performance of the trained model increases with the addition of the two-way interaction between each predictor. To determine whether the interaction is required, an ANOVA test is performed to compare the two trained models. It is extremely statistically significant that the P-value from the anova is less than 2.2e-16. As a result, the second model is chosen to be applied to the testing set to evaluate its performance.

10.5 Results and Discussion

In this study, the multiple linear regression model is applied on the data set during winter from November 2021 to April 2022. A training set is made up of 70% of the randomly chosen data from the gathered data set, while a testing set is made up of 30% of the remaining data. Assessing the performance and correctness of the produced model against already established targets in the collection of predictor variables is the major goal of model testing. The equation of trained model can be written as:

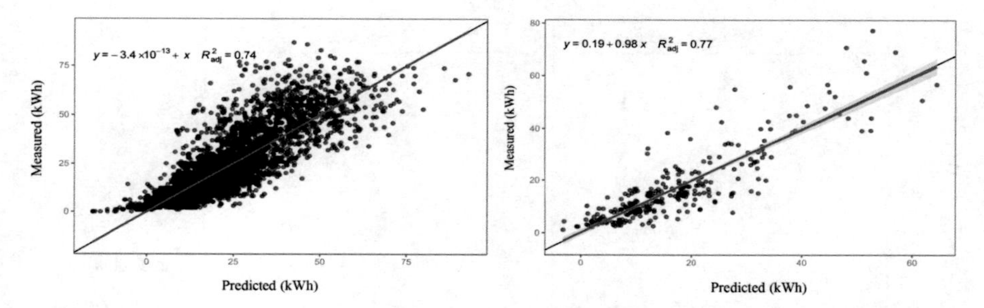

Fig. 10.1 The correlation between measured electricity consumption and predicted electricity consumption using multiple linear regression of two-way interaction with 40 variables: (right) for training set, (left) for testing set

$$
\begin{aligned}
Y &= 0.36 + 0.47x_1 - 0.33x_2 - 0.09x_3 - 0.08x_4 - 0.23x_5 + 0.27x_6 - 0.11x_7 \\
&+ 0.17x_8 + 0.17x_9 - 1.4x_1x_2 - 1.46x_1x_3 + 0.6x_1x_4 + 1.25x_1x_5 - 0.54x_1x_6 \\
&- 0.22x_1x_7 + 0.78x_1x_8 + 0.31x_1x_9 + 0.32x_2x_3 + 0.02x_2x_4 + 0.19x_2x_5 \\
&+ 0.21x_2x_6 - 0.02x_2x_7 - 0.06x_2x_8 + 0.4x_2x_9 - 0.15x_3x_4 - 0.05x_3x_5 + 0.4x_3x_6 \\
&- 0.14x_3x_7 - 0.03x_3x_8 + 0.18x_3x_9 + 0.05x_4x_5 - 0.6x_4x_6 + 0.25x_4x_7 \\
&- 0.03x_4x_8 - 0.08x_4x_9 + 0x_5x_6 + 0.17x_5x_7 - 0.08x_5x_8 - 0.08x_5x_9 - 0.17x_6x_7 \\
&- 0.11x_6x_8 + 0.17x_6x_9 - 0.1x_7x_8 - 0.02x_7x_9 + 0.03x_8x_9
\end{aligned}
$$

This regression model is selected for its highest R^2 value of 0.74 while training. The forecasting between the two sets can be seen in Fig. 10.1. After applying this model on testing set, the R^2 value reached 0.77, higher than the training set.

The regression beta coefficients are represented by the blue line. When the slope for the training set is equal to 1, the regression has the best fit. The calculated regression line does not, however, exactly fit all the data points. The distance between the points and the regression line increases with increasing electricity use. This demonstrates how poorly our model can anticipate the larger values. The comparison of the scaled output by weekday and weekend is shown in Fig. 10.2. The figure compares the values that were measured (in red) with those that the developed model predicted (in sky blue). On weekday during the occupied hours, the predictions are coherent with the measured data up until it exceeds the 30 kWh/h when the predictions begin to underestimate the value. Table 10.2 presents four model evaluation matrices, MAE, MSE, MAPE, and RMSE and together with the model performance R^2. The performance of the model is the best on GMP building and the worst on GEII building, which has R^2 of 55% and MAPE of 69%. That means the interaction between predictors on this building cannot estimate well the energy consumption.

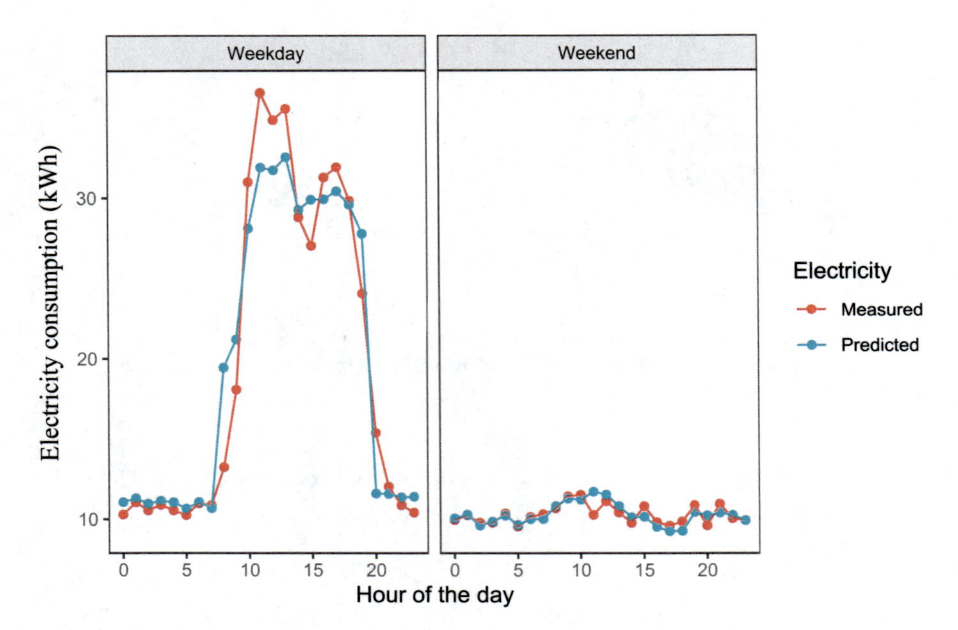

Fig. 10.2 The comparison between measured and predicted electricity consumption hourly over weekday and weekend

Table 10.2 The model errors and performance of each building

Building	MAE (kWh)	MSE (kWh)	MAPE (%)	RMSE (kWh)	R^2(%)
GC	4.33	52.48	0.26	7.24	0.64
GEII	4.76	52.33	0.69	7.23	0.55
GMP	5.44	60.86	0.22	7.80	0.76

10.6 Conclusion

The purpose of this research is to provide a multiple linear regression model that can forecast the hourly electricity usage in educational facilities. To create a one-way and two-way interaction regression model, nine potential explanatory variables were used: CO_2, T_{in}, HR_{in}, T_{ext}, HR_{ext}, SR, day index, hour index, and building area. Better results ($R^2 = 73\%$) are obtained with the combination of two-way interaction models, but the predictor variables also become complex. The model maintains a strong R^2 performance of 74% on the training set and 77% on the testing set. This basic model can be utilized for more research in accordance with Sect. 10.2.1.

The limits happen on bigger values of electricity consumption starting approximately from 30 kWh/h. Moreover, the prediction on GEII building is not acceptable. The preselected predictors might be not the most influential variables. For instance, the outdoor temperature and global solar radiation should be delayed from a few hours to 7 h for a maximum of effect on the indoor temperatures. As this preliminary model has proved itself to be reliable in most cases, a future study will base on this one but adding more potential predictors such as the delayed in indoor temperature, delayed in global solar radiation, and occupancy rate. A validation step should be also included for a further study.

References

1. Anderson, B.: Energy Performance of Buildings Directive, p. 169
2. Deb, C., Zhang, F., Yang, J., Lee, S.E., Shah, K.W.: A review on time series forecasting techniques for building energy consumption. Renew. Sustain. Energy Rev. **74**, 902–924 (Jul. 2017). https://doi.org/10.1016/j.rser.2017.02.085
3. Afroz, Z., Shafiullah, G., Urmee, T., Higgins, G.: Modeling techniques used in building HVAC control systems: A review. Renew. Sustain. Energy Rev. **83**, 64–84 (Mar. 2018). https://doi.org/10.1016/j.rser.2017.10.044
4. Uyanık, G.K., Güler, N.: A study on multiple linear regression analysis. Procedia – Soc. Behav. Sci. **106**, 234–240 (Dec. 2013). https://doi.org/10.1016/j.sbspro.2013.12.027
5. Olive, D.J.: Multiple linear regression. In: Olive, D.J. (ed.) Linear Regression, pp. 17–83. Springer, Cham (2017). https://doi.org/10.1007/978-3-319-55252-1_2
6. Peel, M.C., Finlayson, B.L., McMahon, T.A.: Updated world map of the Köppen-Geiger climate classification. Hydrol. Earth Syst. Sci. **11**(5), 1633–1644 (Oct. 2007). https://doi.org/10.5194/hess-11-1633-2007

Chapter 11
Impact of Urban Morphology on Urban Heat Island Intensity in a Mediterranean City: Global Sensitivity and Uncertainty Analysis

Fatemeh Salehipour Bavarsad, Gianluca Maracchini, Elisa Di Giuseppe, and Marco D'Orazio

11.1 Introduction

Global warming and climate change have wide-ranging effects on the environment as well as on socioeconomic conditions and related sectors, including health. One of the consequences of global warming includes a more frequent occurrence of extreme-weather events such as heat waves (HWs). Urban inhabitants may be more vulnerable to HWs as the UHI effect causes a slower cooling process at night and thus provides little relief from the heat stresses of the day. Considering that 80% of the European population is expected to live in cities within the foreseeable future [1], it is crucial to continue researching the UHI phenomenon and the interaction between global HW and local UHI warming.

For this purpose, urban microclimate simulations (UMsim) are used to simulate urban microclimates. The last decade has seen the development of various simulation tools, whose main differences are related to the spatial scale (i.e., micro-scale, local or neighborhood-scale, and mesoscale) and the resolution of urban environments. Mesoscale models can assimilate information about urban land cover and geometry and may have reliable weather and climate forecasting results. However, due to high computational cost, the extreme complexity of urban areas, and the high number of interacting and uncertain parameters, it is difficult to obtain an accurate representation of the real world. Quick and accurate estimations of UHII can be performed at the neighborhood scale with acceptable computational cost using parametric models such as urban weather generator

F. S. Bavarsad (✉) · E. Di Giuseppe · M. D'Orazio
Università Politecnica delle Marche, Ancona, Italy
e-mail: f.salehipour@pm.univpm.it

G. Maracchini
University of Trento, Trento, Italy

X. Wang (ed.), *Future Energy*, Green Energy and Technology,
https://doi.org/10.1007/978-3-031-33906-6_11

(UWG) [1]. However, the inconsistency between predicted and observed measured values remains a major issue in UMsims. A model's confidence can be raised by obtaining valuable information via sensitivity (SA) and uncertainty (UA) analysis. The relative importance of input parameters can be determined by SA. On the other hand, the effect of input variation on output can be observed through UA [1]. Quite limited numerical studies have employed SA and UA to estimate UHII. For example, the regression technique that Mao et al. [1] used in their estimation of UHII and urban energy consumption in Abu Dhabi, UAE, helped them understand the output variation caused by uncertainty in input parameters. Martinez et al. [2] used the Morris technique to perform SA. Then, they characterized a local microclimate in La Rochelle, France, via UA by UWG. The results highlighted road albedo, vegetation cover, and building height as key parameters. In another study, Kamal et al. [3] performed UA to calculate the impact of building energy load on urban morphology. They discovered that a higher building footprint density could result in more cooling usage. However, an increase in vegetation has little impact on energy saving. To obtain a better understanding of urban thermal features and UHII, reliable data for UMsims are required.

The local climate zone (LCZ) classification approach, a universal classification of urban areas [4], is here adopted to quickly characterize different urban contexts. Also, LCZs contribute to enhancing the accuracy of the contextualization of urban measurements. In this respect, in the present study, we proposed the application of SA and UA to the UWG software to quantitively evaluate the UHII in Rome (Italy) under uncertain urban morphology scenarios (LCZs). Moreover, the relationship between LCZ and UHII is also investigated, which can be a novel aspect of our work.

11.2 Phases, Materials, and Methods

11.2.1 Phases

The methodology section can be divided into three main phases, (I) selecting the most prevalent LCZs in Rome using ArcMap software and WUDAPT database to obtain the urban morphology scenario and (II) establishing the UWG model. In this phase, the range of variation of each LCZ parameter is considered to include in the simulation all the possible scenarios within each LCZ. The building characteristics are also defined based on the literature considering the possible variation of the entire Italian building stock, (III) performing the SA and UA.

11.2.2 The Urban Weather Generator

In the current study, simulations are carried out through the UWG simulation tool to estimate the urban canopy's air temperature profiles and then the UHI effect. The

UWG, previously validated in several studies [2], requires a rural weather file and the characteristics of the selected urban area, including building features. The UWG algorithm includes four major models: (I) the rural station model, (II) the vertical diffusion model, (III) the urban boundary layer model, and (IV) the urban canopy and building energy model. Further detailed information about the conceptual calculation process of the UWG is described in [5]. After each simulation, the average UHII can be estimated as follows:

$$\text{UHII} = \frac{\sum (T_{\text{urb}} - T_{\text{rur}})}{h}, T_{\text{urb}} > T_{\text{rur}} \tag{11.1}$$

where T_{urb} and T_{rur} are, respectively, the urban and rural air temperatures and h is the total number of hours in the simulation.

11.2.3 Sensitivity and Uncertainty Analysis of the UHI Intensity

Methods that rely on SA can be categorized as global and local [1]. The former examines a model's response via simultaneous alteration of all input parameters, while the latter assesses a model's response to a local variation. Since global methods produce more information about the impact of varying model inputs, global models are considered more accurate than local methods. In this research, Sobol's method, which is a global approach based on the analysis of variance, is used to examine the effect of input uncertainty on UHII via the SALib python package [6]. This method breaks up the variance of outputs to the uncertainty of inputs and is a robust method for performing SA. The main outputs are the total-effect indices (ST) and the first order (Si) of the input model parameters. The degree by which the output variance can be reduced when the parameter is fixed is shown by Si. ST is the total-effect Sobol index and considers both the Si and the interaction effect with other parameters. The number of required simulations for the SA is 2 N(1 + D) d, where D is the number of input parameters with uncertainty, while N is the value which must rise until ST values converge. The outcome of these simulations was a matrix used in the UA.

11.2.4 Study Area

The methodology reported in the previous section is applied to the city of Rome, Italy, which is particularly affected by UHI [7] and characterized by very different urban scenarios due to the presence of densely built and low vegetated built areas alternated with large archaeological remains and unbuilt and green area. The city is characterized by a Mediterranean climate (Csa, according to the Köppen-Geiger

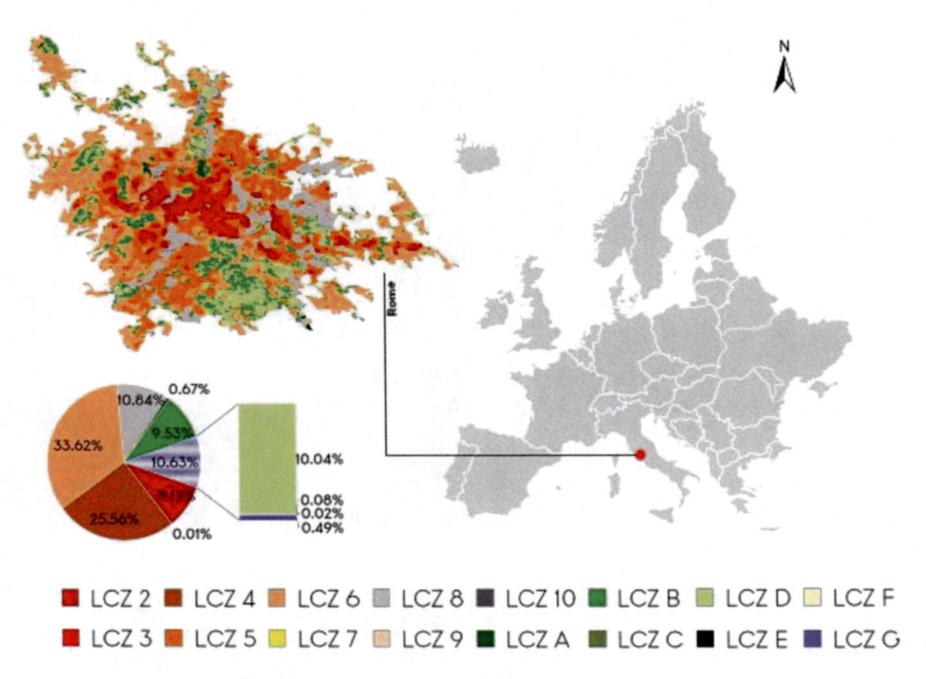

Fig. 11.1 LCZ map of Rome municipality with the percentage of LCZ in this town

classification), characterized by long-hot and dry summers and mild-wet winters [7].

Figure 11.1 reports the share of LCZs in Rome's municipality. In this study, only the most common have been considered in the simulations for the sake of brevity. These are the LCZ2 (9.1%), LCZ5 (25.56%), LCZ 6 (33.62%), and LC8 (10.84%).

The input of the UWG model, which can be classified into urban, rural, and building parameters, are varied within uniform ranges of variation defined based on the relevant literature, as reported in Table 11.1. In particular, the characterization of building features is based on the EU-TABULA project [8] and represents the entire residential building stocks in the Mediterranean areas. UWG parameters such as building density (blddensity), vertical-to-horizontal urban area ratio (vertohor), and grass coverage (grasscover) are computed starting from the LCZ parameters (Table 11.2). All the unreported UWG parameters are maintained in their default values, except for internal gain schedules that are extracted from [9].

Regarding SA, the convergence analysis of ST values is carried out, and the convergence bound for N = 512 is obtained. Considering a total of 28 uncertain parameters, the total number of simulations was 29,696 for each LCZ.

The rural weather files needed for the UWG simulations have been obtained from the Meteonorm software (version 8.0.3) as a typical meteorological year at Fiumicino's airport for 2000–2020. The simulations are carried out in the hottest week of the weather.

Table 11.1 Input parameters of specific LCZs

Parameter [unit]	Description	Distribution (min-max)	Ref.
Urban parameters			
Albveg [−]	Vegetation albedo	0.2–0.3	[1]
Latgrss [−]	Fraction of latent heat absorbed by urban grass	0.45–0.75	[1]
Lattree [−]	Fraction of latent heat absorbed by urban trees	0.5–0.9	[1]
Albroad	Road albedo	0.08–0.24	[1]
BldcooledPerc [%]	The percentage of cooled buildings	0–4	[10, 11]
Sensanth [W/m2]	Sensible anthropogenic heat	5–20	[12]
Treecover [−]	Fraction of urban ground covered in trees	0–1	–
Building parameters			
U-value [W/m^{2}K]	Window U-value including film coefficient	1.8–5.7	[13]
Glzr [−]	Glazing ratio	0.11–0.31	[9, 14]
SHGC [−]	Solar heat gain coefficient	0.2–0.8	[1]
N_occ [m^2/person]	Occupancy density	15–25	[1]
Infil [ACH]	Infiltration rate	0.1–0.83	[13, 14]
SetpointCool [°C]	Cooling set point	22–26	[13]
COP [−]	COP of cooling system	2.5–3	[13]
Albroof [−]	Albedo of roof	0.25–0.6	[15]
Albwall [−]	Albedo of wall	0.25–0.6	[15]
Q-light [W/m2]	Maximum light process load per unit area	1.6–11.5	[13]
U-value [W/m-K]	Wall thermal transmittance	0.23–2.58	[13]
U-value [W/m-K]	Roof thermal transmittance	0.33–3.44	[13]
U-value [W/m-K]	Floor thermal transmittance	0.34–2.9	[13]

Table 11.2 Input parameters for each LCZ. Bld_sf: Building surface fraction; prev_sf: previous surface fraction; H/W: canyon aspect ratio; bld_height: building height [4]

LCZ	**LCZ of Rome**	**Description**	**bld_sf [%]**	**perv_sf [%]**	**H/W [-]**	**Mean Height [m]**
LCZ$_2$		Compact mid-rise	40-70	0-20	0.75-2	10-25
LCZ$_5$		Open mid-rise	20-40	20-40	0.3-0.75	10-25
LCZ$_6$		Open low-rise	20-40	20-50	0.3-0.75	3-10
LCZ$_8$		Large low-rise	30-50	0-20	0.1-0.3	3-10

LCZ map of Rome municipality with the percentage of LCZ in this town file, starting from August 14.

11.3 Results and Discussion

11.3.1 Uncertainty Analysis of Urban Microclimate

The results of UA are presented and discussed in this section. Figure 11.2 illustrates the envelope's urban temperature profile based on hourly maximum (upper bound) and minimum (lower bound) temperature compared with rural temperature profiles (solid line) for the selected LCZs in Rome during the simulation period. In general, the graph verifies a very slight UHI between the late morning and early afternoon. Conversely, UHII values in nighttime (20:00–06:00) are more pronounced than in daytime (05:00–19:00). These upward trends are observed in all selected LCZs. Previous studies proved that UHI is mainly a nighttime phenomenon since urban materials absorb solar energy, and accumulated heat is dissipated into the urban canyon at night [16]. In particular, the average daytime hourly maximum values of UHII are 0.99, 0.82, 0.76, and 0.88 °C in LCZ2, LCZ5, LCZ6, and LCZ8, respectively. Also, the average nighttime hourly maximum values of UHII are 4.1 °C, 3.7 °C, 3.6 °C, and 3.5 °C in LCZ2, LCZ5, LCZ6, and LCZ8, respectively. These results are encountered before in the literature on UHI [17], and the temperature of 5 °C is recorded as a variation of the average maximum UHII in a dense urban area in Rome [17]. The urban temperature in LCZ2 is higher than the other LCZs because LCZ2 is the densest district among all considered LCZs and may lead to a higher amount of heat being released from air conditioning into the urban canopy (see Table 11.2).

The results of the UA have been reported in Fig. 11.3a in terms of average UHII for each LCZ (see Eq. 11.1). As expected, the higher average of UHII is found

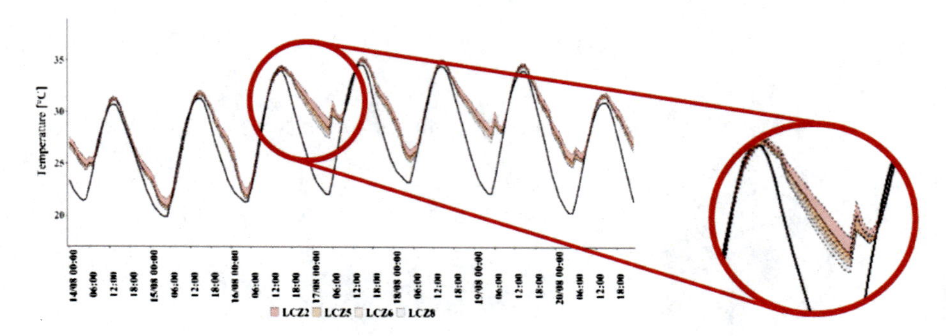

Fig. 11.2 The envelope's urban air temperature profiles for different LCZs and their comparison with rural temperatures (solid line) during the simulation periods in Rome

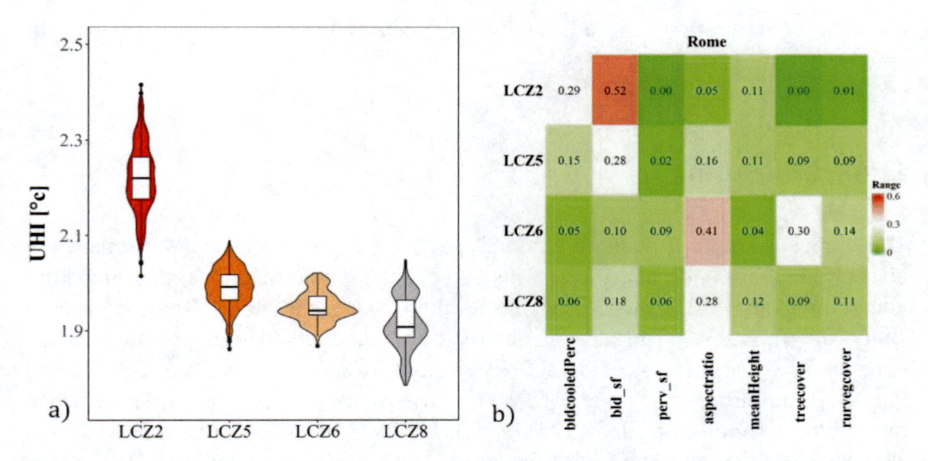

Fig. 11.3 (**a**) Violin plots with box-whisker plots of UHII; (**b**) ST obtained for the most important parameters in each LCZ

in LCZ2, followed by LCZ5, LCZ6, and LCZ8. In particular, the average UHII in LCZ2, LCZ5, LCZ6, and LCZ8 is 2.2, 1.99, 1.94, and 1.91 °C, respectively. These results are compatible with the UHII reported in the literature. For example, in the study carried out by Bonacquisti et al., the average UHII in Rome is about 3 °C. Previous studies also indicated a substantial variation of UHI across various LCZs, particularly in built areas [18]. In another study [19], the authors found out that among analyzed LCZs, LCZ6 had the lowest UHI intensity compared to LCZ1, which had the highest one.

11.3.2 Sensitivity Analysis

Figure 11.3b indicates the ST results obtained from Sobol's SA for each LCZ, which represents a ranking of parameters based on the influence of their variance on the output variance. Additionally, this graph reports only the most influential parameters among all inputs with the highest value of ST. Since STs are expressed in a relative term, a comparison between LCZ can be made in terms of input ranking, from the most important (red) to the least important (green). It is foreseeable that morphological parameters have the highest impact on average UHII. The fraction of land surface covered with buildings (bld_sf) is the most influential parameter in LCZ2 and LCZ5, and also it is the most influential one in LCZ8, followed by buildings that used an air conditioning system (bldcooledPerc), which is one of the most in LCZ2 because this urban zone has compact building distribution and a high number of buildings [20]. Besides, the aspect ratio is the most influential parameter in LCZ6 and LCZ8 and the second key parameter in LCZ5. The tree coverage (treecover) is the second key parameter in LCZ6, which probably depends on urban configurations and a high rate of greenery in these LCZs rather than in other selected LCZs.

11.4 Conclusion

This paper presents a detailed SA and UA in the UWG model to estimate the UHI effect under uncertainty scenarios and evaluate the impact of urban modeling uncertainty on UHII. Considering the Mediterranean climate of Rome as a case study, the variations in urban characteristics are applied to four prevalent urban contexts in Rome (i.e., LCZ2, LCZ5, LCZ6, and LCZ8). The results of UA showed that UHII is stronger in LCZ2 (2.2 °C), where the district has a dense urban morphology, followed by LCZ5 (1.99 °C), LCZ6 (1.94 °C), and LCZ8 (1.91 °C). These results are consistent with the results obtained in other studies. SA results highlight that the urban morphology parameters (LCZ parameters) such as building surface fraction, aspect ratio, and mean height of buildings are the most influential ones in this study. However, it cannot be assuredly suggested that a "lower" LCZ always corresponds to a higher UHII in a given city because of the overlap between the values of UHII in the different LCZs and should be more investigated in various towns and climates. In perspective, these results can help to understand the relationships between regional urban morphology and the magnitude of UHI at the district scale. Besides, it would be useful for sustainable urban planning and building design to mitigate UHI problems in future urban growth. It is recommended that the influence of urban characteristics in the study of urban climate is considered to adopt the appropriate mitigation action.

References

1. J. Mao, J. H. Yang, A. Afshari, L. K. Norford, 'Global sensitivity analysis of an urban microclimate system under uncertainty: design and case study', Build. Environ., vol. 124, pp. 153–170, Nov. 2017, https://doi.org/10.1016/j.buildenv.2017.08.011
2. Martinez, S., Machard, A., Pellegrino, A., Touili, K., Servant, L., Bozonnet, E.: A practical approach to the evaluation of local urban overheating– a coastal city case-study. Energ. Buildings. **253**, 111522 (2021). https://doi.org/10.1016/j.enbuild.2021.111522
3. Kamal, A., et al.: Impact of urban morphology on urban microclimate and building energy loads. Energ. Buildings. **253**, 111499 (2021). https://doi.org/10.1016/j.enbuild.2021.111499
4. World Urban Database and Access Portal Tools, 'World Urban Database'.: http://www.wudapt.org/create-lcz-training-areas/step-2/ (2015)
5. Kim, H., Gu, D., Kim, H.Y.: Effects of Urban heat Island mitigation in various climate zones in the United States. Sustain. Cities Soc. **41**(January), 841–852 (2018). https://doi.org/10.1016/j.scs.2018.06.021
6. Herman, J., Usher, W.: SALib: sensitivity analysis library in python (Numpy). contains sobol, SALib: an open-source Python library for sensitivity analysis. J. Open Source Softw. 2(9), 97 (2017). https://doi.org/10.1016/S0010-1

7. Salata, F., Golasi, I., de Lieto Vollaro, R., de Lieto Vollaro, A.: Outdoor thermal comfort in the Mediterranean area. A transversal study in Rome, Italy. Build. Environ. **96**, 46–61 (2016). https://doi.org/10.1016/j.buildenv.2015.11.023
8. TABULA webtool, 'No Title', [Online]. Available: https://webtool.building-typology.eu/#bm
9. Lavagna, M., et al.: Benchmarks for environmental impact of housing in Europe: definition of archetypes and LCA of the residential building stock. Build. Environ. **145**(May), 260–275 (2018). https://doi.org/10.1016/j.buildenv.2018.09.008
10. entranze, 'No Title'.: https://entranze.enerdata.net/share-of-dwellings-with-air-conditioning.html. (2010)
11. Eurostat.: https://ec.europa.eu/eurostat
12. Manatsa, D., Chingombe, W., Matarira, C.H.: The impact of the positive Indian Ocean dipole on Zimbabwe droughts tropical climate is understood to be dominated. Int. J. Climatol. **2029**, 2011–2029 (2008). https://doi.org/10.1002/joc
13. Carnieletto, L., et al.: Italian prototype building models for urban scale building performance simulation. Build. Environ. **192**, 107590 (2021). https://doi.org/10.1016/j.buildenv.2021.107590
14. Mata, É., Sasic Kalagasidis, A., Johnsson, F.: Building-stock aggregation through archetype buildings: France, Germany, Spain and the UK. Build. Environ. **81**, 270–282 (2014). https://doi.org/10.1016/j.buildenv.2014.06.013
15. G. Maracchini, A. Latini, E. Di Giuseppe: Un nuovo strumento per analisi di incertezza e sensibilità su strategie di mitigazione del fenomeno Isola di Calore Urbana, pp. 16–19. (2021)
16. Bavarsad, F.S., Di Giuseppe, E., D'Orazio, M.: Numerical assessment of the impact of roof albedo and thermal resistance on urban overheating: a case study in southern Italy. Sustain. Energy Build. **2022**, 125–134 (2021)
17. Salvati, A., Monti, P., Coch Roura, H., Cecere, C.: Climatic performance of urban textures: analysis tools for a Mediterranean urban context. Energ. Buildings. **185**, 162–179 (2019). https://doi.org/10.1016/j.enbuild.2018.12.024
18. Hidalgo García, D., Arco Díaz, J.: Modeling of the urban Heat Island on local climatic zones of a city using sentinel 3 images: urban determining factors. Urban Clim. **37** (2021). https://doi.org/10.1016/j.uclim.2021.100840
19. Liu, Y., Li, Q., Yang, L., Mu, K., Zhang, M., Liu, J.: Urban heat Island effects of various urban morphologies under regional climate conditions. Sci. Total Environ. **743** (2020). https://doi.org/10.1016/j.scitotenv.2020.140589
20. Maracchini, G., Bavarsad, F.S., Di Giuseppe, E., D'Orazio, M.: Sensitivity and uncertainty analysis on urban Heat Island intensity using the local climate zone (LCZ) schema: the case study of Athens BT – sustainability. Energ. Buildings, 281–290 (2022/2023). https://doi.org/10.1007/978-981-19-8769-4_27

Part III
Power System and Electric Power Load Management

Chapter 12
Self-Adaptive Ageing Models for Optimal Management and Planning of Assets in Microgrids

Thierry Coosemans, Wouter Parys, Cedric De Cauwer, Maitane Berecibar, and Maarten Messagie

12.1 Introduction

The evolution towards decentralization of energy production leads towards a large increase of new and flexible devices in low- and medium-voltage smart grid environments, often within the context of a microgrid or an energy community. Each of these assets has its own degradation curve, failure liability and dispatching opportunities. In this type of environment, an advanced, dynamic and autonomous asset management system will be essential to ensure grid reliability, sustainability targets and economic optimization. Key in such system would be the availability of a self-adaptive ageing modelling algorithms using real historical operational data of assets to assess the future capacity, remaining useful life and optimal operational profiles. As such investments and operations can be forecasted, planned and optimized in the short and long term in microgrids ranging from a low complexity level up to highly critical and fail-save systems. Asset degradation (and thus behaviour change) is typically slow in comparison with operational state changes, and ageing algorithms hence be carried out in an off-line environment using low sampling rates. This paper will propose an ageing modelling methodology based on machine learning techniques using historical data of assets operated in real-life circumstances and with limited knowledge asset properties, e.g. size and age. The methodology uses several regression methods going from a simple linear regression towards more advanced neural network regression techniques, as well as clustering techniques (K-means and spectral clustering). Until now our work mainly focused on applying the method on Battery Energy Storage Systems (BESS) and solar installations (PV), but it will be extended to a larger range of assets. The

T. Coosemans · W. Parys (✉) · C. De Cauwer · M. Berecibar · M. Messagie
MOBI Research Centre, Vrije Universiteit Brussel, Brussels, Belgium
e-mail: Thierry.Coosemans@vub.be; Wouter.Parys@vub.be; Cedric.De.Cauwer@vub.be; Maarten.Messagie@vub.be; Maitane.Berecibar@vub.be

X. Wang (ed.), *Future Energy*, Green Energy and Technology,
https://doi.org/10.1007/978-3-031-33906-6_12

work described is part of the H2020 BD4OPEM project funded by the European Commission (Grant agreement ID: 872525). This project aims to create an open energy marketplace that offers innovative AI-based services, to enable the efficient management of energy distribution grids and associated assets. The authors wish to thank the European Commission for its support.

12.2 Methodology

12.2.1 State of the Art

Throughout literature, a wide spectrum of different approaches and methodologies are found to model the degradation of different microgrid assets. These could be roughly divided in two categories:

- Pre-trained deterministic models: These models are typically built upon testing data in a controlled environment.
- Self-adaptive algorithms that include the measured operational for adapting the model parameters.

The concept of the second category is pursued during the development of the methodology. The main idea is that the method will work with a minimal amount of background information and the maximal amount of available operational data of the asset itself. An overview of self-adaptive algorithms for the battery asset is described in [1] and can be subdivided in the following categories (an overview is shown in Fig. 12.1):

- Deterministic models that adapt itself by re-fitting or retraining when the available data is updated: ARIMA, SVR, GPR, RVM and ANN [2–5].
- Dynamic model with updating parameters of a parametric model using filtering techniques: Bayesian particle filter models [6–8].

In comparison with the wide range of researched state of health and degradation models, only a very limited number of hands-on software applications are available that generate useful outputs for microgrid operators, developers and investors. This counts especially for tools that consider historical operational profiles. Current available software tools on the market are typically developed for batteries; examples are BLAST [9], which simulates ageing based on typical operational properties, Argonne [10], Monte Carlo curve fitting technique for cell-to-cell manufacturing variations and internal component ageing and DNV [11] and degradation estimation based on state-of-the-art battery degradation models derived from independent cell testing data. Comparable but also very limited tools exist for other assets. However, all examined existing tools fail to deliver the combination of all the functionalities and results that are needed for optimizing energy scheduling, configuring lifetime extension measures and performing a coherent investment planning over the

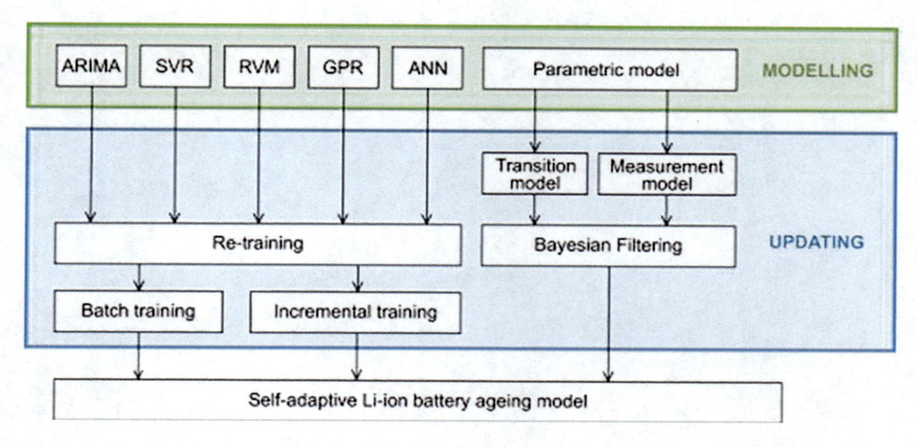

Fig. 12.1 Overview of self-adaptive (battery) models [1]

different assets based on operational data only. These missing features are the focus of this methodology:

- Built model upon own asset historical data (without access to state-of-the art technology-specific ageing models).
- Update model periodically.
- Derive degradation cost for different operational regimes.
- Calculate current operational limitations and properties.

12.2.2 Approach of the Methodology

The aim of this methodology is to assess the operational limitations and the ageing of a wide range of assets that are potentially part of a smart low-voltage distribution grid or microgrid and with focus on the assets that have a considerable property change over its lifetime of the asset. However, the developed algorithms will not only analyse the state of health parameter (=parameter that changes over time). Also, the current operational limitations and properties are closely monitored as they are typically valuable inputs for an energy management system (e.g. charging limits at high state of charge for batteries, irradiation to power conversion in different seasons for PV installations, minimal and maximal load for heat pumps, etc.) The methodology is fully developed for Battery Energy Storage Systems (BESS) and solar installations (PV); however the methodology and building blocks are developed in a way that they are easy adaptable to incorporate other asset types as well. The next paragraphs will illustrate the approach based on the more complex case of a battery, followed by a brief description of the methodology applied on PV installations.

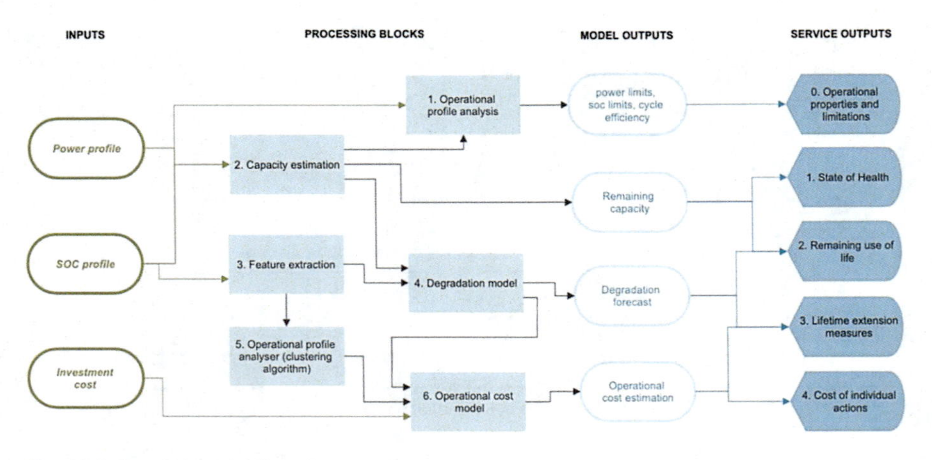

Fig. 12.2 Battery model flow diagram

Batteries The method is built up in a modular way where each processing block builds upon historical time series data of the asset, on the output of other processing blocks or on a combination of both. Figure 12.2 shows the complete service data and processing flow for the battery system. Hereafter, the building blocks of the model for the battery system are explained in detail.

Processing Block 1: Operational Profile Analysis The first processing block is a simple data analysing block that calculates the current operational properties of the battery encompassing, power limitations for different state of charge levels, minimal and maximal state of charge and cycle efficiency. These parameters are typically nameplate properties of the battery asset or BMS input settings. However, the theoretic value is not always corresponding with reality and can deviate depending on the operational circumstances.

Processing Block 2: Capacity Estimation The methodology followed in this processing block focuses on the typical available data for the different microgrid assets. In the case of a battery, this means the power profile and state of charge profile. This processing block is a purely analytical one, based on the properties of the battery asset. The algorithm estimates the battery charging and discharging capacity based on a basic coulomb counting methodology and extracting the full capacity based on the equation:

$$\text{capacity} = \frac{\sum_{t}^{t+n} (\text{power} * \Delta t)}{\Delta \text{SOC}}$$

This results in a raw capacity estimation over time, which is then smoothed and filtered to generate the final estimated battery capacity degradation curve over time.

Processing Block 3: Feature Extraction In the prospect of building a degradation model for the asset, the idea is to find the correlation between the operational profile and the capacity degradation curve. However, training a model on real time series data will not generate an accurate degradation model and would not allow one to create new and unseen operational regimes. In fact, it is key to transform the time series into features that are closely related to the principles behind the battery degradation. In this way the model will explore the chosen features during training. Typical features encompass total energy in/total energy out, total amount of cycles, total time at constant state of charge, nominal average power (dis)charging, depth of discharge and state of charge level. However, to fully capture the degradation properties, a cross-link between obvious features, such as (high) charging (power) at high state of charge, or (high) discharging (power) at low state of charge, could be even more relevant. For some models (e.g. linear regression model), it is important to explicitly calculate the cross-linked features as well to create the training dataset. More complex non-linear models (e.g. multi-layered neural networks, tree regressors) are able to connect different features independently. Figure 12.3 shows the results of the feature extracting algorithm for some examples of daily state of charge profiles.

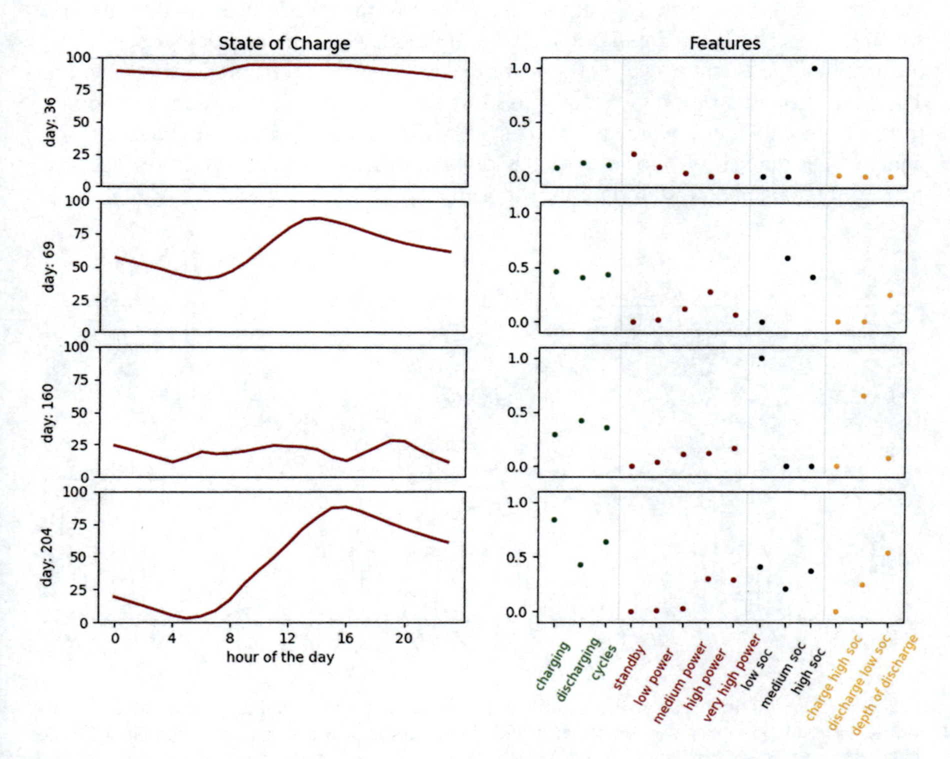

Fig. 12.3 Visualization of the feature extracting algorithm: transformation from a daily operational profile to a normalized feature space

Processing Block 4: Degradation Modelling This processing block holds a machine learning model to link operational profiles to induced degradation. Because of the slow degradation of a battery (multiple months to see a capacity decrease), it does not make sense to link a daily profile to daily degradation when training the model. This makes the model unstable and induces overfitting as daily variations and measurements errors are taken fully into account during the training phase. To avoid this problem, the following rules are followed:

- Choose a long enough period to combine to extract a relevant degradation step.
- Choose stable and relevant features.
- Calculate the average time series features over the full period to train the model.
- Use a moving window method to create enough samples (every sample of multiple dates is only a 1-day shift from the previous sample).

For every new dataset coming in, a grid search is performed over different machine learning regression models, the length of the time series combined in one training sample and the amount and type of features to include in the model.

Processing Block 5: Operational Profile Analysing This extracts typical operational regimes by using clustering algorithms (e.g. K-means, K-nearest neighbours, spectral clustering). The input of the clustering model are typically daily operational profiles that are transformed by the feature extraction model of processing block 3. The output of this block is directly used as input for the degradation cost model (processing block 6), whereas the cost of different operational regimes can be analysed. Figure 12.4 shows how the daily profiles are transformed to specific battery features and clustered by the algorithm.

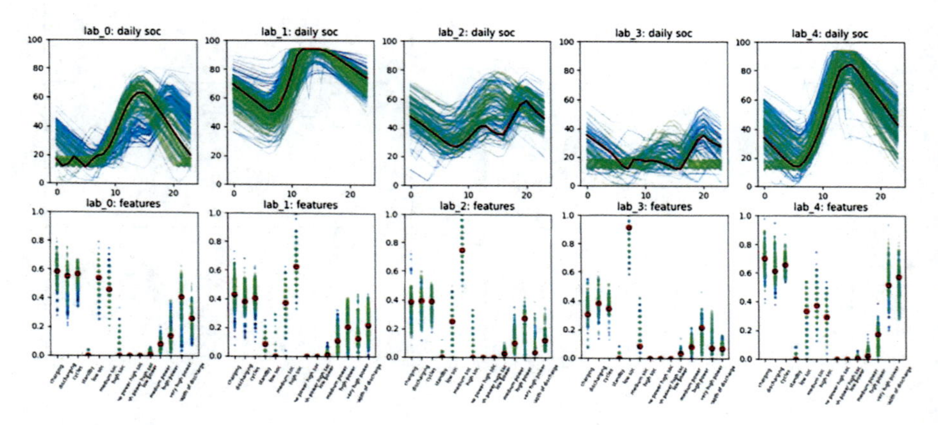

Fig. 12.4 The top figures show the resulting clustered daily profiles, and the bottom figures show the corresponding battery features for each individual profile as they are used as the input for the clustering and degradation algorithm

Processing Block 6: Operational Cost Model The operational cost processing block considers the investment cost and uses the output of the degradation model and the operational profile analysis to calculate the specific degradation cost for the different operational profiles. The core idea is solely to amortize the investment cost over the lifetime of the battery based on the effective operations.

Outputs of the Modelling The outcome of the modelling results in the estimation of operational parameters such as power limitations, state of charge limitations and cycle efficiency, state of health, remaining use of life and cost of individual actions.

PV Systems As PV installations are non-dispatchable assets, no operator impact and user variations are considered over time. The input is solely the production profile and relevant weather data (temperature and irradiance). Processing blocks 3, 5 and 6 of Fig. 12.2 are discarded as they have to do with different operational regimes and operational costs. The state of health feature that is modelled is the current peak production capacity of the installation (which is known to degrade slowly over the lifetime of the PV installation). The degradation model does not incorporate the operational profiles as the degradation is only a function of time. The degradation trend will be extrapolated to calculate the remaining useful life.

12.3 Results and Conclusions

12.3.1 Results

The methodology was applied on a real dataset of a large-size PV installation at the VUB Hospital in Brussels and on a dataset created by a simulation of a battery in a multi-asset microgrid environment with a smart battery control.

PV System The used PV installation does have long-term data availability, whereas production is linked with irradiation to determine the operational parameters, the state of health and the degradation forecast. The decay of two operational properties was determined; the conversion of solar irradiation to electric power was found to decay 0.8% per year, and the maximal power output throughout the year was found to decrease 0.9% yearly, indicating a remaining capacity of 92.9% after 8 years of operation. This corresponds with the typical manufacturer's warranty of 90% remaining capacity after 10 years of operation. A summary of the results can be seen in Fig. 12.5.

Battery System The used battery data is produced by a long-term simulation of a microgrid with a varying load profile, a varying energy price, a pv system and an energy management system controlling the battery in a smart way. The battery ageing behaviour and states are simulated in a complex way, considering depth of discharge, (dis)charging current rates, state of charge, battery temperature and battery age. The resulting battery behaviour is a simplification of a real battery in a

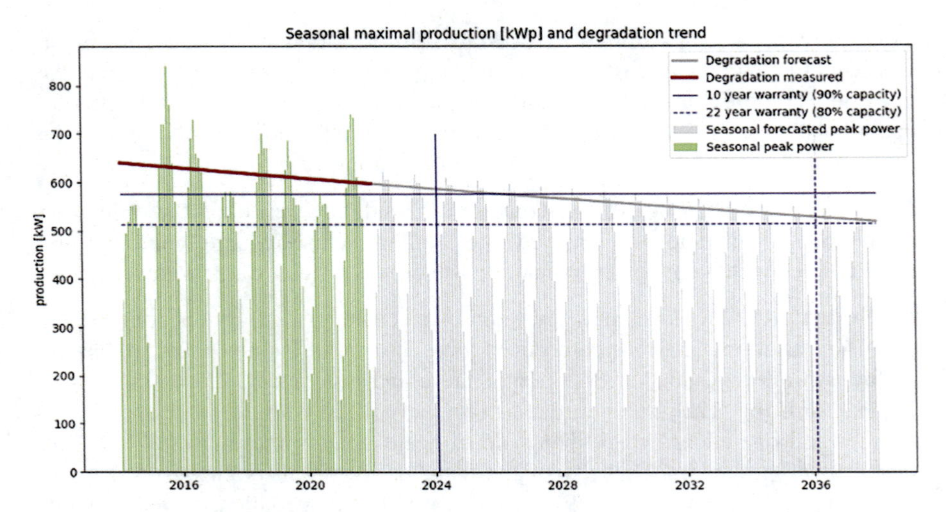

Fig. 12.5 Historical degradation and future degradation prognosis in visualized in combination with the production capacity warranty

real application. However, the goal of this research step is to validate the possibility of a machine learning model to learn the dynamic behaviour of battery ageing in online applications with solely easy measurable operational parameters without prior knowledge of battery-specific ageing models nor the need for battery-specific ageing tests.

Figure 12.6 shows the results for a 28-month simulated battery operation (in a temperature-controlled area). The first 7 months are used for training the model, and the 21 following months are used for testing the model. The operational profile changes over the seasons and over the years to test the model's adaptability to unseen circumstances. For training the model, each 40 consecutive days of state of charge profiles are aggregated before performing the feature extraction algorithm. Four different regression models are applied.

As the model is trained by aggregating the battery behaviour of 40 consecutive days, it also forecasts degradation on a 40-day-based average. The long-term prediction results thus in a smooth curve following the seasonal trends of low battery stress in winter and high battery stress in summer. The model accuracy lies in between 95.4% and 98.1% for the full battery capacity decrease forecast over a period of 21 months.

As discussed in Sect. 12.2.2 (processing block 6), in a secondary phase, the trained models are used to create a feedback loop to the smart battery control system for short-term daily operational choices. Figure 12.7 shows the results of the degradation cost forecast (by means of predicted capacity loss of the battery) of the trained models for different potential short-term operational regimes. The output of the degradation model will result in an improved energy management system as the predicted battery degradation cost can be an important new decision parameter.

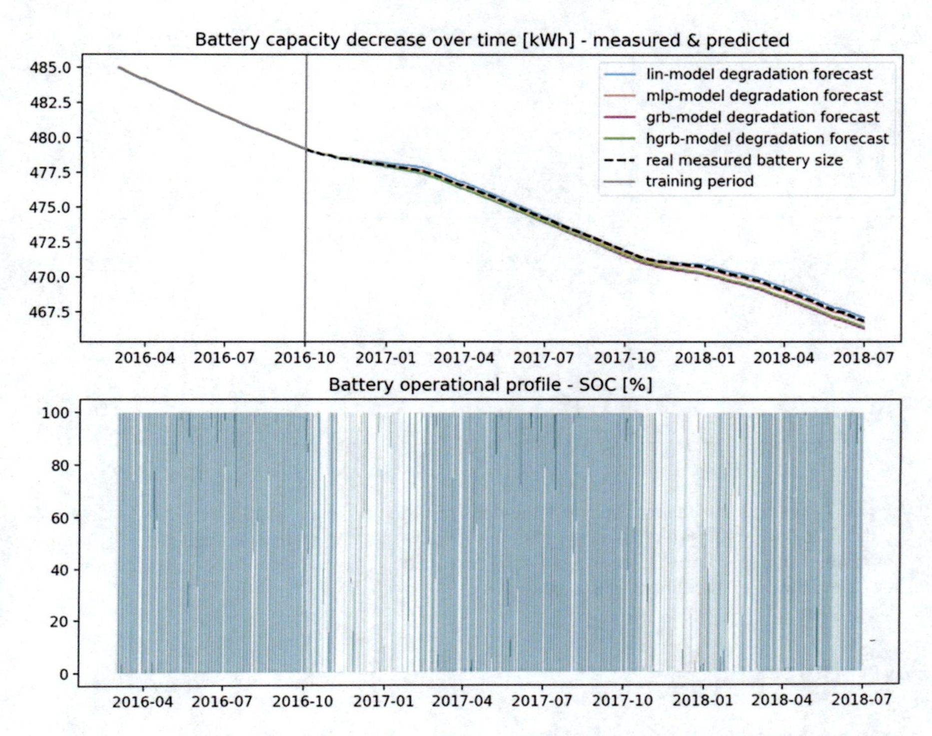

Fig. 12.6 Historical and future degradation of the battery in relation to an imposed operational profile. Four different types of machine learning models are applied on the dataset. (*lin* linear regression, *mlp* feed forward neural network regression, *grb* gradient boosting regression, *hgrb* histogram-based gradient boosting regression)

Model type	lin	mlp	grb	hgrb	**Real value**
Start capacity (month 7) [kWh]	479,17	479,17	479,17	479,17	**479,17**
End capacity (month 28) [kWh]	467,07	466,58	466,40	466,27	**466,838**
Absolute model error [kWh]	0,233	0,263	0,437	0,567	/
Absolute model error [%]	0,05%	0,06%	0,09%	0,12%	/
Total capacity decrease [kWh]	12,10	12,60	12,77	12,90	**12,332**
Total capacity decrease [%]	2,52%	2,63%	2,66%	2,69%	**2,57%**
Relative model error [%]	1,89%	2,13%	3,54%	4,60%	/

12.3.2 Conclusions and Future Work

A self-adaptive ageing model based on historical operational data for optimal management and planning of assets was developed and validated by experimental data for PV systems and simulated data for battery systems. Further work on short term will focus on further validation of the methodology for batteries on experimental data of real battery applications.

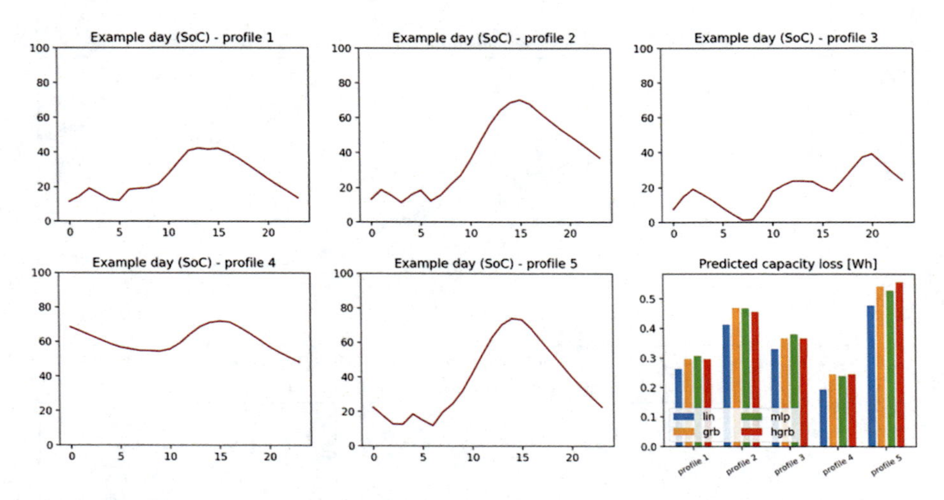

Fig. 12.7 Five typical battery use profiles and the resulting capacity loss prediction for each profile by the four different types of machine learning models that are applied on the dataset. (*lin* linear regression, *mlp* feed forward neural network regression, *grb* gradient boosting regression, *hgrb* histogram-based gradient boosting regression)

References

1. Lucu, M., Martinez-Laserna, E., Gandiaga, I., Camblong, H.: A critical review on self-adaptive Li-ion battery ageing models. Power Sources. **401**, 85–101 (2018). https://doi.org/10.1016/j.jpowsour.2018.08.064. https://www.sciencedirect.com/science/article/pii/S0378775318309297
2. Long, B., Xian, W., Jiang, L., Liu, Z.: an improved autoregressive model by particle swarm optimization for prognostics of lithium-ion batteries. Microelectron. Reliab. **53**, 821–831 (2013). https://doi.org/10.1016/j.microrel.2013.01.006
3. Tipping, M., Mach, J.: Sparse Bayesian learning and the relevance vector machine. Learn. Res. **1**, 211–244 (2001). https://doi.org/10.1162/15324430152748236
4. Liu, D., Pang, J., Zhou, J., Peng, Y., Pecht, M.: Prognostics for state of health estimation of lithium-ion batteries based on combination Gaussian process functional regression. Microelectron. Reliab. **53**, 832–839 (2013). https://doi.org/10.1016/j.microrel.2013.03.010
5. Liu, J., Saxena, A., Goebel, K., Saha, B., Wang, W.: An adaptive recurrent neural network for remaining useful life prediction of lithium-ion batteries. In: Annual Conference on Prognostics and Health Management Society, Portland (2010)
6. Xing, Y., Ma, E.W.M., Tsui, K.-L., Pecht, M.: An ensemble model for predicting the remaining useful performance of lithium-ion batteries. Microelectron. Reliab. **53**, 811–820 (2013). https://doi.org/10.1016/j.microrel.2012.12.003
7. Wang, D., Yang, F., Tsui, K., Zhou, Q., Bae, S.J.: Remaining useful life prediction of lithium-ion batteries based on spherical cubature particle filter. IEEE Trans. Instrum. Meas. **65**, 1282–1291 (2016). https://doi.org/10.1109/TIM.2016.2534258
8. Yang, F., Wang, D., Xing, Y., Tsui, K.-L.: Prognostics of Li(NiMnCo)O_2-based lithium-ion batteries using a novel battery degradation model. Microelectron. Reliab. **70**, 70–78 (2017). https://doi.org/10.1016/j.microrel.2017.02.002

9. BLAST – Battery Lifetime Anlaysis and Simulation Tool Suite. https://www.nrel.gov/transportation/blast.html
10. Argonne – Battery Life Estimator. https://www.anl.gov/partnerships/battery-life-estimator
11. DNVGL – Battery degradation assessment and warranty evaluation. https://www.dnv.com/services/battery-degradation-assessment-and-warranty-evaluation-159269

Chapter 13
Study on Fast Charging Using Phase Change Materials for Electric Vehicle Applications

Maitane Berecibar, Hamidreza Behi, and Theodoros Kalogiannis

13.1 Introduction

The energy crisis that we are facing is pushing even more the electric mobility. The increase of usage of electric vehicles (EV) is happening in any kind of road transportation, buses, trucks, passenger cars, bicycles, etc. The technology in place is safe and ready to fulfill the user needs; however enabling fast charging is still a pending action for EVs. Therefore, research is being done in all the battery-related topics: novel materials, smart functionalities, cell designs, smart charging schedules, novel cooling/heating systems, and Battery Thermal Management System (BTMS) strategies to take higher fast charging rates.

During the recent years, numerous efforts have been made to design BTMS solutions that can meet the needs of e-mobility with regard to mileage and range anxiety. To efficiently address the high-heat dissipation rates, in [1] we evaluated the phase change material (PCM) assisted with heat pipe cooling system for thermal management of lithium titanium oxide (LTO) battery cells. Hence, a passive cooling solution with natural convection is experimentally tested and validated with numerical models. Results showed that a significant reduction in the temperature rise at the fast C rates can be achieved with approximately 40% improvement compared to solely natural convection of the cells. Also, studies on the optimization of the thermal management design tools have shown that improvements on the performance during a fast charge can be achieved while efficient thermomechanical analysis is taken under consideration [2, 3]. Results in previous studies have shown that a co-design automated optimization study can support up to 15%, 70%, and 40% improvement on the maximum temperature, cell uniformity, and cell-level heat

M. Berecibar · H. Behi (✉) · T. Kalogiannis
Vrije Universiteit Brussel, Brussels, Belgium
e-mail: maitane.berecibar@vub.be; hamidreza.behi@vub.be; theodoros.Kalogiannis@vub.be

X. Wang (ed.), *Future Energy*, Green Energy and Technology,
https://doi.org/10.1007/978-3-031-33906-6_13

distribution, respectively, on the proposed BTMS, with a simultaneous 5% reduction on the overall volume of the battery pack, compared to its baseline design.

Although the discovery of novel materials will help in overcoming this bottleneck, the developed work aims to provide solutions to the EV systems that are now ready to be used. Therefore, the focus is to enhance the BTMS strategies using novel PCM in the cooling/heating system. It is the aim of the BTMS to keep that battery temperature at the optimal temperature, which is around 25 °C [4]. However, due to performance and degradation of the battery, it is difficult to keep the battery temperature constant. Therefore, and due to safety concerns, the battery systems need to have on board a thermal management system and an appropriate cooling/heating system. There are multiple ways to develop the cooling/heating systems, air based, liquid based (water or refrigerant), or hybrid systems. However, now the use of PCM materials is under research to evaluate potential solutions [5].

13.1.1 SELFIE

This paperwork is based on the ongoing European Project SELFIE. This project is funded from the H2020 program. SELFIE stands for "SELF-sustained and Smart Battery Thermal Management SolutIon for Battery Electric Vehicles." The aim of SELFIE is to develop a novel, smart, energy-efficient, and modular BTMS capable and flexible of allowing further capacity and voltage to reach longer distances. SELFIE targets fast charging solutions and aims to test a 5 C-rate charging rate by reducing the charging time to 10 min by absorbing the excess heat created in the fast charging process. Novel components and control schemes are being developed to ensure not only optimal operation of the battery cells but also passenger comfort by controlling at the same time both operations at any ambient conditions. The total system will be integrated and validated in a Fiat Doblo vehicle [6].

13.1.2 Cell Type

Within the lithium-ion technology, there are numerous battery chemistries that can be used. In the SELFIE project, after a selection process, nickel manganese oxide (NMC) cathode and lithium titanium oxide (LTO) anode type of cell was selected [7]. This type of cell was selected due to its advantages such as fast charging capability, high power density, and larger cycle life, making this cell technology suitable for high-power mobility applications. The cell specifications are described in Table 13.1.

Table 13.1 Table captions should be placed above the tables

Parameter	Value
Chemistry	LTO
Shape	Prismatic
Nominal voltage (V)	2.3
Maximum voltage (V)	2.7
Minimum voltage (V)	1.5
Capacity (Ah)	23
Specific energy (Wh/kg)	96
Energy density (Wh/L)	203
Weight (kg)	0.550
Volume (L)	0.260
Dimensions L × W × H (mm)	115 × 22 × 103
Heat-specific capacity (J/kg.K)	1150
Thermal conductivity x, y, z (W/m.K)	31, 0.8, 31

13.2 Modeling

13.2.1 Numerical Model

As a first step, the electrical model of the described cell is developed. The model is based on a dual-polarization electrical equivalent model (Fig. 13.1), and it consists an open-circuit voltage source (Voc), a resistance connected in series (R0), and two R//C branches connected in parallel representing the time-dependent polarization processes that take place at an applied potential to the battery terminals (Vcell). The parameters have been obtained by a set of characterization experiments. Within those experimental tests, the preconditioning, the discharge capacity and the HPPC tests are developed at different current rates. For this modeling purpose, both Matlab/Simulink and Comsol Multiphysics tools are used. Finally, the model validation is done through the testing of the world harmonized light vehicle test procedure (WLTC) profile. The design of experiments and the parameterization process have been performed in our previous study presented in [8]. The model has been validated with direct current pulses and electrochemical impedance spectroscopy (EIS) at various charge/discharge rates and ambient conditions. Various parameterization methods have been studied on the effects of computational power, modeling accuracy, and robustness, while the impedance output is utilized as the irreversible Joule losses input for the 3D Comsol thermal model.

13.2.2 Experimental Setup

The experimental setup was done at module level, wherein 30 cells were connected in series. In this case, no space was placed between the cells. Additionally, for more

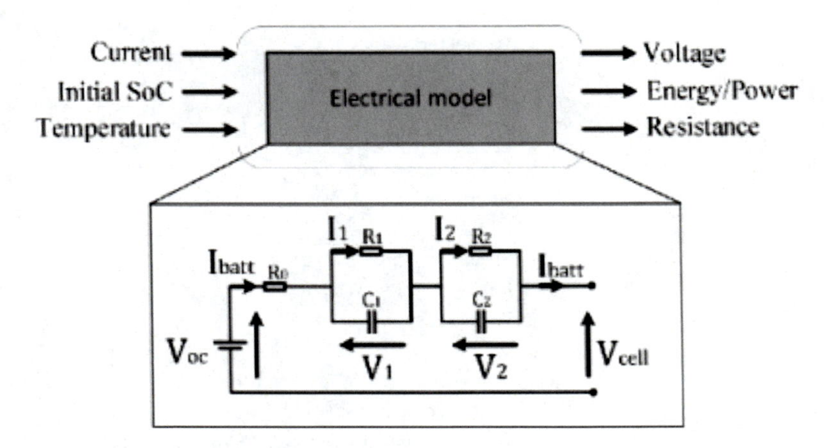

Fig. 13.1 Dual-polarization electrical equivalent model

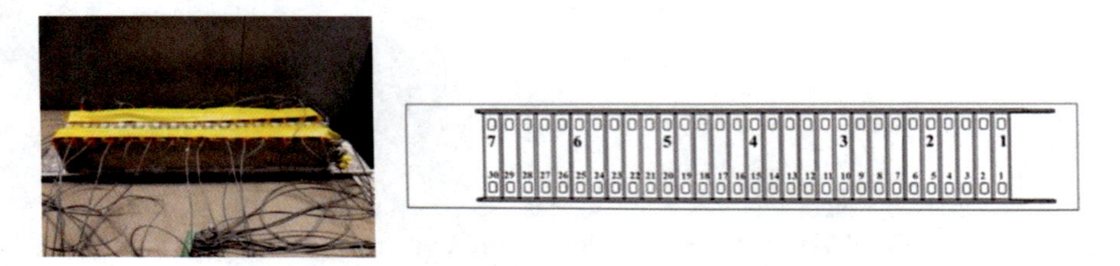

Fig. 13.2 Module development (left) experimental setup (right) thermocouple location

reliable and secure connections, aluminum fins were welded to each of the cell tabs. Copper busbars were used for cell connection. The developed setup is shown in Fig. 13.2, where on the right side a picture of the module can be seen and on the left side the experimental setup sketch and the location of the thermocouples is shown. The properties of the developed battery module, the PCM, and cooling liquid can be found here [9]. The experimental test setup consists of the SELFIE battery module, the electrical control with the commercial battery management system (BMS), an Accel chiller to control the inlet temperature of the coolant and its mass flow rates, a PEC battery cycler that is capable of delivering 80 V and 600A, a PICO acquisition system for the data collection based on K-type thermocouples, and a personal laptop to monitor and collect final data. The input power of the module is controlled by the PEC software interface. As shown in the Fig. 13.2, the 30S1P 69V module is tested with various loading profiles to evaluate the effect of PCM and liquid cooling at the ambient temperature. Seven thermocouples are placed on the cells as shown in the right-hand figure at the top-side of the cells, which location is selected as it is closer to the cell tabs, and it has been measured to produce the highest heat generation [1].

13.3 Results

13.3.1 Validation

The numerical model is validated with a fast charging experimental study at 4C. In the experimental setup, a hybrid cooling system is used and the temperature of a certain cell is measured using the thermocouples of K-type. In Fig. 13.3 the validation results are shown, confirming that the results of the experiment and the model are in good agreement. It is shown that the proposed hybrid cooling can preserve the battery pack temperature within the recommended safe zone, without provoking any hazardous scenario. The PCM heat buffer placed on the bottom of the cells acts as a good support on the heat dissipation. Cell temperature at location 5 is shown here, which is more away from the coolant inlets. We observe that a good agreement between the experimental and numerical solutions is achieved with the cell temperature at the end of a 5C charge being below 33 °C.

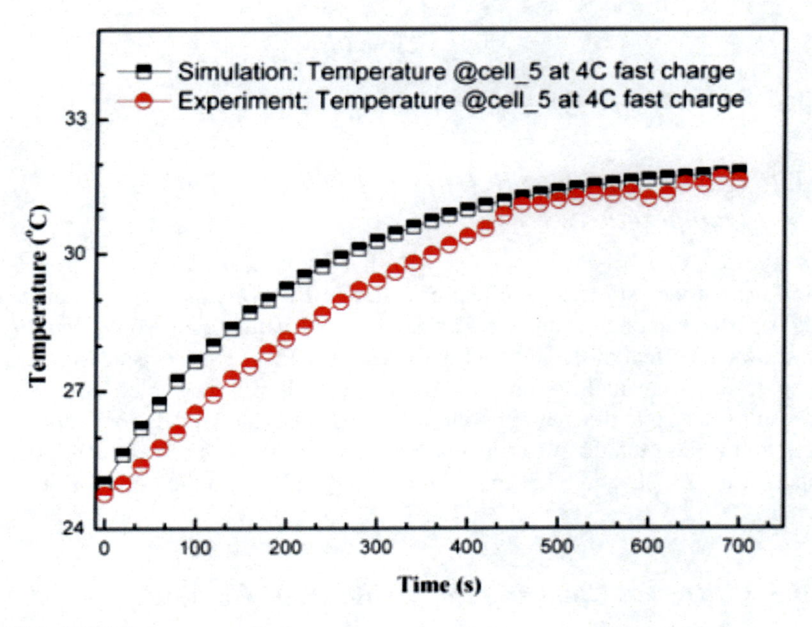

Fig. 13.3 Validation

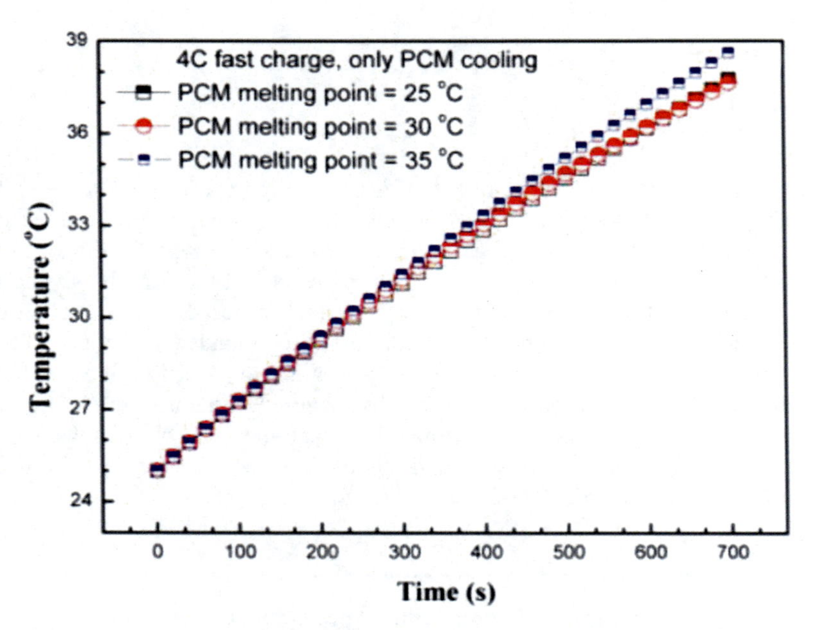

Fig. 13.4 Effect of PCM melting temperature

13.3.2 Fast Charging with Different PCM Melting Temperatures

In this section we estimate the effect of different PCM melting temperatures. The study is performed on the numerical model, once the validation is confirmed. Figure 13.4 shows the effect of the temperature when only PCM is used as cooling system strategy at different melting points. The liquid cooling is neglected in this study. The result shows that the melting point of the PCM has a relatively low impact with a variation of less than 2 °C at the end of the simulations. The results at different temperatures, 25, 30, and 35 °C, are shown in Fig. 13.4 with the same result.

13.3.3 Different Charging Rates with PCM Cooling

The C rate has a severe impact on the battery temperature. Therefore, different C rates have been tested to see the effect of the PCM cooling. Figure 13.5 shows the reached temperatures at 2C, 4C, and 6C where 34.8 °C, 38.6 °C, and 43.6 °C have been respectively reached. The results confirm that using only PCM cooling is not sufficient.

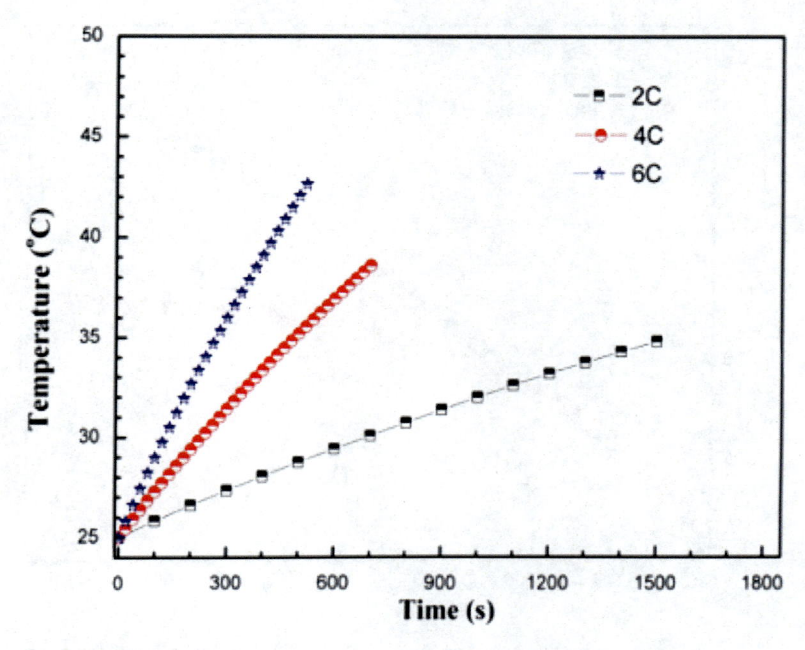

Fig. 13.5 The battery module tested at different C rates: 2C, 4C, and 6C

13.3.4 Fast Charging with Hybrid Cooling

In the next test, a comparison between the usage of only PCM as cooling or combining it with liquid cooling is shown. Figure 13.6 presents the result of this test developed at 4 C charging rate. The red line corresponding to the combination of liquid cooling and PCM shows better results that only PCM cooling (black line), reaching 38.4 °C only with PCM and 31.8 °C with the hybrid cooling.

Moreover, there are also differences with the temperature uniformity. Figure 13.7 shows those differences in both cases, by using PCM or hybrid cooling. It is observed that local hot spots are developed on the cells if they are only cooled from the passive PCM solution. These can affect their lifetime capabilities, as localized degradation modes can be generated and contribute to the loss of capacity of increase of internal resistance [10]. In the latter case, a cooling solution should be built by taking into consideration the higher heat dissipation rates over the applications lifetime.

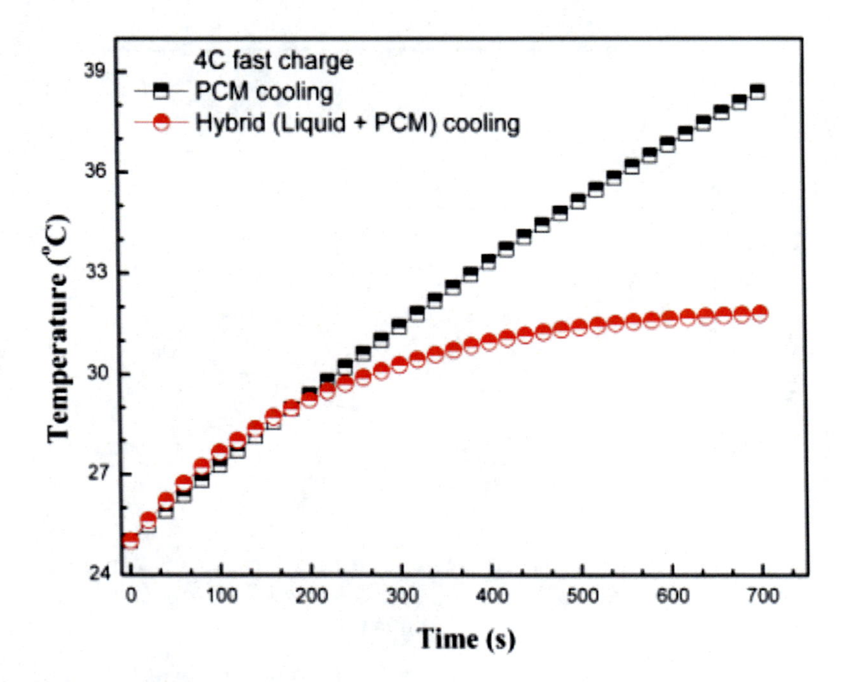

Fig. 13.6 Comparison of PCM and hybrid cooling

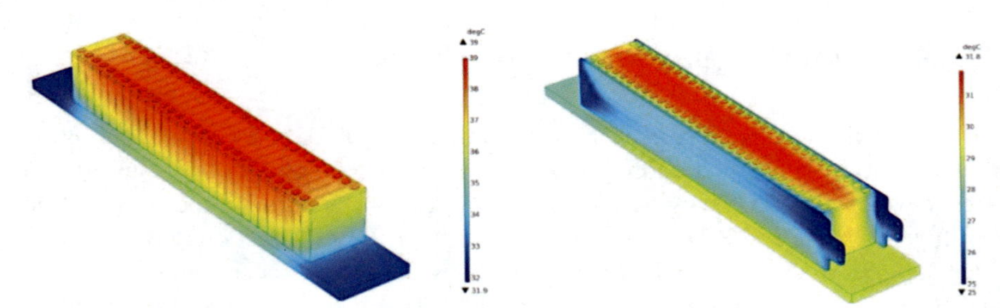

Fig. 13.7 Temperature uniformity with PCM (left) with hybrid cooling (right)

13.4 Conclusions and Future Work

In this paper we presented the study of a self-sustained battery module that is developed for the SELFIE project. The study investigates the passive and active cooling solution in parallel operation, as an attempt to meet the heat requirements of the cells during a 10-min complete charge. The tested results show that PCM cooling alone as a passive solution is not sufficient for fast charging scenarios. As drawbacks of PCM, one should highlight the low impact on the melting temperature at the battery temperature. Therefore, for fast charging or ultrafast charging scenarios, a hybrid cooling is needed by utilizing fluid coolants. The showed results cover the needs of the scenario by using a water/glycol-based liquid coolant. Future work is ongoing by extending the model to pack level and optimizing the flow rate of the tested coolant and evaluating the robustness of the system on the whole battery pack, implemented in the battery vehicle.

Acknowledgments The presented study was developed under the structure of the SELFIE project that was granted from the European Union's Horizon 2020 research and innovation program under Grant Agreement Nr. 824290.

References

1. Behi, H., Karimi, D., Heidari, F., Akbarzadeh, M., Khaleghi, S., Kalogiannis, T., Hosen Md, S., Jaguemont, J., Van Mierlo, J., Berecibar, M.: PCM assisted heat pipe cooling system for the thermal management of an LTO cell for high-current profiles. Case Stud. Thermal Eng. **25**(2021), 100920 (2021)
2. Kalogiannis, T., Hosen Md, S., Heidari, F., Akbarzadeh, M., Jaguemont, J., Van Mierlo, J., Berecibar, M.: Multi-objective particle swarm optimization and training of datasheet-based load dependent lithium-ion voltage models. Electr. Power Energy Syst. **133**(2021), 107312 (2021)
3. Kalogiannis, T., Akbarzadeh, M., Hosen Md, S., Jaguemont, J., Behi, H., De Sutter, L., Lu, J., Van Mierlo, J., Berecibar, M.: Effects analysis on energy density optimization and thermal efficiency enhancement of the air-cooled Li-ion battery modules. J. Energy Storage. **48**(2022), 103847 (2022)
4. Behi, H., Karimi, D., Youssef, R., Patil, M.S., Van Mierlo, J., Berecibar, M.: Comprehensive passive thermal management systems for electric vehicles. Energies. **14**, 3881 (2021)
5. Behi, H., Karimi, D., Jaguemont, J., Gandoman, F.H., Khaleghi, S., Van Mierlo, J., Berecibar, M.: AEIT International Conference of Electrical and Electronic Technologies for Automotive (AEIT AUTOMOTIVE), Turin (2020)
6. SELFIE Project, European Union's Horizon 2020 research and innovation programme under grant agreement No 824290, https://eu-project-selfie.eu
7. Gonzalez-Aguirre, E., Gastelurrutia, J., Patil, M.S., Portillo-Valdes, L.: Avoiding thermal issues during fast charging starting with proper cell selection criteria. J. Electrochem. Soc. **168**, 110523 (2021)
8. Kalogiannis, T., Hosen Md, S., Akbarzadeh, M., Goutam, S., Jaguemont, J., Lu, J., Qiao, G., Berecibar, M., Van Mierlo, J.: Comparative study on parameter identification methods for dual-polarization lithium-ion equivalent circuit model. Energies 2019. **12**, 4031 (2019). https://doi.org/10.3390/en12214031
9. Behi, H., Karimi, D., Kalogiannis, T., Jiacheng He, Patil, M.S., Muller, J.D., Haider, A., Van Mierlo, J., Berecibar, M.: Advanced hybrid thermal management system for LTO battery module under fast charging. Case Stud. Thermal Eng. **33**, 101938 (2022)
10. Heidari, F., Jaguemont, J., Goutam, S., Gopalakrishnan, R., Firouz, Y., Kalogiannis, T., Omar, N., Van Mierlo, J.: Concept of reliability and safety assessment of lithium-ion batteries in electric vehicles: basics, progress, and challenges. Appl. Energy. **251**, 113343 (2019)

Chapter 14
Impact of Smart Hydronic System with Heat Pump on Electricity Load of a Typical Queensland Household

Adrian Rapucha, Ramadas Narayanan, and Meena Jha

14.1 Introduction

14.1.1 Challenges of the Australian Electricity Grid

A rapid transition towards renewable energy sources (RES) can be observed in many regions of the world. Australia is a world leader in this process, reaching the highest annual capacity increase of new renewable energy per citizen [1]. According to the Australian Energy Market Operator (AEMO), the country will reach 50% of annual electricity generation from RES in 2025 and 90% in the mid-2030s [2]. Undoubtedly, it will be a remarkable achievement; however, the national grid already struggles with problems related to the high penetration of RES. Photovoltaic (PV) and wind energy, which have the largest share of RES in Australia, are dependent on weather conditions. Their output is out of the grid operator's control and often does not align with the current demand. It makes balancing demand and supply levels much more challenging compared to entirely fossil fuel-based generation. Moreover, the grid infrastructure developed over decades was designed as a one-way system, providing customers with electricity from centralised power plants. Due to the high number of small decentralised PV installations, this route is periodically disturbed, and surplus electricity is sent from households and businesses back to the grid. This may lead to low voltage grid straining [3]. Due to these factors, Australian National Electricity Market (NEM)

A. Rapucha (✉) · M. Jha
School of Engineering and Technology, Central Queensland University, Sydney, NSW, Australia

R. Narayanan
School of Engineering and Technology, Central Queensland University, Sydney, NSW, Australia

University Drive, Bundaberg, QLD, Australia
e-mail: r.narayanan@cqu.edu.au

X. Wang (ed.), *Future Energy*, Green Energy and Technology,
https://doi.org/10.1007/978-3-031-33906-6_14

has significantly decreased stability in the last years, resulting in frequency and voltage fluctuations and numbers of blackouts [4]. Therefore, further renewable energy integration requires solutions that can ensure the stability and reliability of the power system.

14.1.2 Impact of Domestic Air Conditioning and Water Heating on the Grid

Air conditioning is a basic system for many houses in Australia. The number of AC devices has been increasing in recent years, and it is expected that, soon, 80% of Australian households will be equipped with at least one air conditioner [5]. According to one of the studies, even up to 83% of these devices operate simultaneously during extremely hot days [6]. Considering the current climate changes and the fact that we can expect an increase in the number of these extremely hot days and overall annual cooling demand, the air conditioning electricity load, especially peak load, will be rising. Moreover, the recent pandemic raised discussion about appropriate ventilation of buildings. Increasing ventilation air flow would cause additional heating, cooling and dehumidification load, therefore, more sustainable solutions are necessary to decrease electricity consumption [7]. Residential AC is one of the major contributors to the greatest peak demand events and can account for 25% of its load [8]. In addition, the household's space conditioning load significantly changes during the day [9]. This variability caused by occupants' behaviour patterns significantly contributes to destabilising grids. Domestic Hot Water (DHW) preparation, the second-most energy-consuming process in Australian houses, also intensifies the negative residential sector impact as the highest hot water demand occurs simultaneously with peak electricity loads [10].

Therefore, domestic heating and cooling is an objective of many demand response programs and studies. Nevertheless, despite these programs which have already been introduced (e.g. Peak Smart Air Conditioners in Queensland) definitely help to manage the grid, their ability to stabilise the network is very limited. It is due to the low inertia of the conventional AC system. Air has low heat capacity. Therefore, turning-off AC to save energy when the grid is overloaded will result in a quick drop or increase in indoor temperatures and occupants' inconvenience. Thus, most demand response programs are mostly restricted to only tempering AC capacity during the highest peak loads and using pre-cooling and pre-heating periods before expected overloads.

14.1.3 Hydronic Systems with Heat Pumps in Grid Balancing

Heat pumps, like air conditioners, use a refrigeration system to provide cooling and heating. However, the major difference is that heat pumps use an additional hydronic system that distributes thermal energy in a building. Hydronic systems can also be easily adapted to produce DHW. Thermal energy storage (TES) tanks, due to the high volumetric capacity of water, can be used as buffers that absorb energy during periods of high RES generation and deliver it to space during peak hours [11] without disturbing the occupant's preferences. The thermal mass of the building can additionally increase the storage capacity and inertia of the system [12], for example, slabs with hydronic underfloor heating. These properties allow for the application of hydronic systems and heat pumps in smart grids. This is confirmed by numerous studies [11–14]. Moreover, many heat pump manufacturers have already introduced Smart Grid Ready (SGR) devices, oriented whether on increasing self-consumption of household PV generation, reacting to grid operator commands, or both. Despite these facts, due to the still low penetration of heat pumps in Australia, this topic has not been adequately investigated for this market before. Only a few publications evaluated their potential to increase the self-consumption of household rooftop PV generation [15–17]. However, they did not include measured air conditioning demand. The measured electricity load of heating and cooling devices contains not only information about building heating and cooling loads but also provides insight into the behaviour patterns of occupants.

Considering the mentioned problems and the potential of a hydronic system with heat pump and smart management, it was decided to evaluate the possible impact of such a system on domestic electricity load if it would replace the conventional Australian system consisting of AC and electric water heater in a typical Queensland household. Therefore, a series of TRNSYS simulations of heat pump energy consumption under standard and SMART control strategies have been conducted. Heating and cooling loads for the hydronic system have been created from measured annual electricity consumption data of an AC system in a typical Queensland household. Then, the results of the simulations were compared to the measured input data using various criteria.

14.2 Study

14.2.1 Input Data

A major input data for simulations, which was cooling and heating load for hydronic system, was created from the data provided by Commonwealth Scientific and Industrial Research Organisation (CSIRO) which shows the bulk 30 minutes data of electricity consumption of various circuits, including AC, measured in dozens of Queensland households in the study conducted between 2012 and 2017 [18]. Since

2016 was the last full year of the available data, this year has been for analysis. To estimate produced thermal energy Q by AC, knowing only its electricity load P used the below equation:

$$Q = COP_T \bullet P \tag{14.1}$$

COP_T is a COP of an air conditioner at a specific ambient temperature. COP_T was determined using performance data of one of the popular middle-range air conditioner series [19]. Ambient temperature data came from one of the Brisbane weather stations. It is assumed that the days with mean ambient temperature <18 °C are heating days, and ≥ 18 °C are cooling days. The modelled hydronic system also provides domestic hot water. Therefore, 30 minutes usage profile for hot water has been prepared using average measured data from 5 Australian households [20].

14.2.2 TRNSYS Model

TRNSYS models of hydronic systems based on air source heat pump (ASHP) and ground source (GSHP) consisted of 6 major components: heat pump, DHW Tank, Chilled Water thermal energy storage (TES), Hot Water TES, control and circuits imposing heating, cooling and DHW loads. The model has implemented performance data of GSHP and ASHP that are available in Australian or New Zealand markets and are typically used in residential applications. Their capacities have been adjusted to match the maximum cooling load, which is dominant in Queensland. DHW tank has a volume of 200 litres and is set to model average models available on the market. The flow mixing valves are added to ensure that consumed hot water always has 50 °C, to meet Australian requirements. TES tanks have the same thermal loss properties as the DHW tank. The volumes were determined using a formula that applies low inertia systems that are based on various manufacturers' recommendations and is dependent on maximum cooling or heating load $Q_{\max}$ of the building [21]:

$$V_{TES} = 81.54 + 53.8 \bullet Q_{\max} \tag{14.2}$$

The results have been rounded to sizes typically available in the market. For heating, TES has a volume of 200 litres, while cooling TES has a volume of 300 litres. Then, heating and cooling loads are imposed by a circuit with auxiliary heating and cooling elements that discharge TESs according to the prepared earlier load profiles.

The standard control strategy is similar to the strategy used in a typical residential system. DHW heating is a priority, with tank set temperature at 57 °C and 6 °C hysteresis. Heating and cooling set temperatures are determined using a weather compensation curve. It means that tanks temperatures are set depending on the ambient conditions. This option is commonly used in heat pump systems to increase their efficiency [13]. For the standard control strategy, the minimum water temperature for cooling is 7 °C, and the maximum is 18 °C. In heating mode, the minimum water temperature is 30 °C and the maximum 45 °C. Hysteresis for cooling and heating is 2 °C.

The first step of creating a SMART control strategy was the determination of low and high electricity demand periods. For this purpose, the historical 30 minutes wholesale electricity price for QLD shared by AEMO [22], which indicates the supply and demand ratio, has been used. Prices close to the annual average suggest a well-balanced grid. A high price is a sign of increased demand or insufficient generation. A low-price period occurs when supply exceeds consumption. The price in $/MWh has been divided into five categories according to the percentage difference from the mean price $\Delta\%$. Results are presented in Table 14.1.

These categories have been used to adjust the set temperatures to promote energy consumption during periods of categories 1 and 2 and limit energy usage for categories 4 and 5. The adjustment has been conducted by a formula that changes the weather compensation values for cooling and heating. For categories 1 and 2, the temperatures have been changed, causing increasing available thermal energy in tanks. Category 1 is characterised by quicker and greater adjustments. If before category 3 system prepared energy reserve, it will be maintained for 2 hours and then slowly discharged. For the rest of the time, Category 3 system has the same settings as for Standard control. For Categories 4 and 5 system limits the energy consumption using stored energy. For the SMART system, hysteresis has been decreased to 1 °C to allow a quicker reaction.

The preparation of DHW also has been adjusted in the SMART control strategy. DHW tank temperature set is not fixed but changes according to the electricity price category, as presented in Table 14.2.

14.3 Results and Discussion

Results of TRNSYS simulations allowed for comparison of energy consumption of the conventional hydronic with a standard control strategy and hydronic with a SMART control strategy systems. The first analysis criterium was annual energy consumption, presented in Table 14.3.

All hydronic systems allowed for a significant reduction in annual energy consumption. This is mostly due to the replacement of an electric resistive DHW heater with the refrigeration system with an average annual COP for this task of about 3.6 for ASHP and 3.0 for GSHP. ASHP in cooling mode used slightly more energy than conventional air conditioning. However, GSHP reached a better result than AC for this task regardless of the control strategy. Heating was more energy efficient using a hydronic system than air conditioning in all simulated cases. Usage of the SMART control strategy caused an increase in energy consumption compared to the standard control strategy. It is caused by forcing the heat pump to work in conditions in which it is less efficient, as well as increased heat losses.

Table 14.1 Categories of wholesale electricity price

Δ%	< −50%	>= −50%	< −25%	>= −25%	Mean	< 25%	>= 25%	< 50%	>= 50%
Price	$33.72	$33.72	$50.58	$50.58	$67.44	$84.30	$84.30	$101.16	$101.16
Category	**1**	**2**		**3**			**4**		**5**

Table 14.2 DHW set temperature compensation for the SMART control strategy

Category 1	Category 2	Category 3	Category 4	Category 5
+4 °C	+1 °C	–	–5 °C	−7 °C

Table 14.3 Annual total and per task energy consumption

	DHW	Cooling	Heating	Total
System	*kWh*	*kWh*	*kWh*	*kWh*
Conventional	930.6	574.0	132.3	1637.0
ASHP standard	330.3	601.7	104.3	1036.4
ASHP SMART	356.8	628.8	113.0	1098.5
GSHP standard	392.3	493.1	103.5	988.9
GSHP SMART	423.3	522.9	107.6	1053.8

Table 14.4 Distribution of electricity consumption during each price category

Price category	1		2		3		4		5	
System	*kWh*	%	*kWh*	%	*kWh*	%	*kWh*	%	*kWh*	%
Conventional	279.9	17.1	504.6	30.8	528.5	32.3	105.2	6.4	218.8	13.4
ASHP standard	151.3	14.6	317.7	30.7	343.8	33.2	77.3	7.5	146.3	14.1
ASHP SMART	366.0	33.3	320.2	29.1	320.2	29.1	34.6	3.2	57.5	5.2
GSHP standard	156.5	15.8	305.9	30.9	311.8	31.5	67.8	6.9	146.9	14.9
GSHP SMART	376.1	35.7	311.3	29.5	288.1	27.3	30.7	2.9	47.7	4.5

More interesting outcomes appear when electricity usage is analysed in terms of the time of consumption. Table 14.4 shows the consumption of all systems during each of the five electricity price categories.

Hydronic systems with a standard control strategy and the conventional system reached similar proportions in each category, and the difference does not exceed 2.5 percentage points. It means that these systems would still contribute to grid straining and peak loads. However, because they use less energy than the conventional system, the peak loads could be somewhat decreased. Nevertheless, that would also mean a decrease in consumption during the lowest category, making overgeneration even more problematic to manage. SMART control strategy, which utilises the hydronic system storage benefits, allowed for shifting most of the space conditioning and water heating load from Categories 4 and 5 to Category 1. It means that the SMART hydronic system not only significantly reduces the household load during the peaks but also consumes surplus energy generated by RES. Systems under the SMART control strategy consumed even more energy during Category 1 than the AC and electric water heater. Figure 14.1 presents 30 minutes of electricity usage of conventional and hydronic SMART systems as well as price during the sample day of the analysed year. It can be seen how the SMART heat pump system reacts to the variable price and uses valleys to charge tanks that are then used during peaks.

The unveiled results show great potential for hydronic systems with SMART management in balancing the Australian grid. Loads of air conditioning and water heating, which are the most energy-consuming residential processes, can be shifted

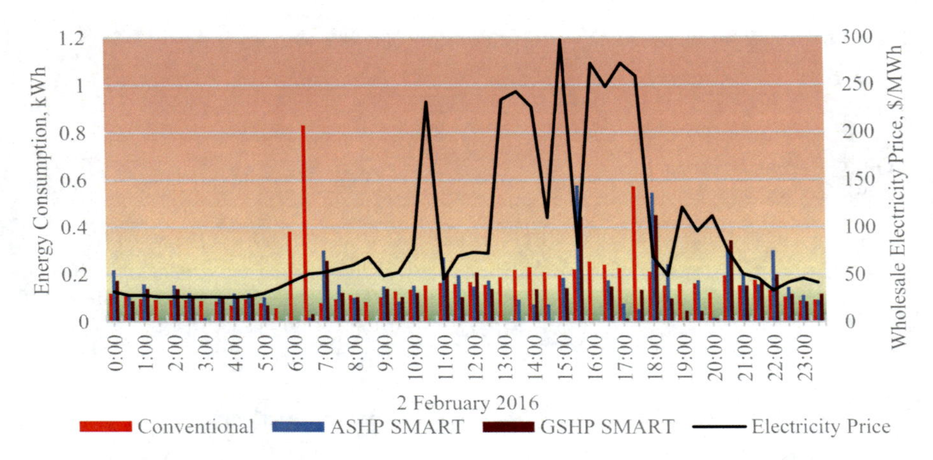

Fig. 14.1 30 minutes electricity price and consumption profile of SMART and Conventional systems for 02.02.2016

from high demand to low demand periods to a great extent using these types of systems. It is also worth mentioning that hydronic systems achieved an almost unnoticeable risk of only small under conditioning at about 0.5% time of the year. Taking into consideration the amount of load that has been shifted, this is a huge benefit of demand response systems based on heat pumps over AC. To reach the same effect, a supply and demand driven response program for AC would cause much greater disturbance in the comfort level of the occupants. That would make these types of systems unwanted by the users.

14.4 Conclusions

Among many considered tools that may help to maintain the stability of the Australian grid, demand response programs for the residential sector are one of the most important. Considering increasing problems with peak loads, which are highly driven by domestic space conditioning, and constantly growing RES penetration, including small household installations, grid operators will need the development of storage capacity and electricity consumption management. Both of them can be provided by hydronic systems and heat pumps. As presented in this paper, heat pumps with smart management can not only decrease the energy consumption of domestic space conditioning and hot water preparation but also can be extremely useful in shifting this load from periods with high demand to times of surplus generation. The simulations of the hydronic system under SMART control management that adjust heat pump output conducted for a typical Queensland household showed that it is possible to reach a reduction of electricity consumption during the highest peak loads (Category 5) by 73.7% for ASHP and 78.2% for GSHP comparing to the conventional system. Most of the load is realised during Category 1, increasing the proportion of energy consumption for this category from 17.1% for the conventional solution to 33.3% for ASHP and 35.7% for GSHP. Therefore, it can be concluded that hydronic systems based on heat pumps with SMART management have great potential to stabilise the Australian grid.

References

1. Blakers, A., Stocks, M., Lu, B., Cheng, C., Stocks, R.: Pathway to 100% renewable electricity. IEEE J. Photovolt. **9**(6), 1828–1833 (2019)
2. Australian Energy Market Operator AEMO.: 2022 integrated system plan. Version 1.0 (2022)
3. Currie, G., Evans, R., Duffield, C., Mareels, I.: Policy options to regulate PV in low voltage grids-Australian case with international implications. Technol. Econ. Smart Grids Sustain. Energy. **4**, 1–10 (2019)
4. Arraño-Vargas, F., Shen, Z., Jiang, S., Flether, J., Konstantinou, G.: Challenges and mitigation measures in power systems with high share of renewables—the Australian experience. Energies. **15**(2), 429 (2022)
5. Goldworthy, M., Poruschi, L.: Air-conditioning in low income households; a comparison of ownership, use, energy consumption and indoor comfort in Australia. Energ. Build. **203**, 109411 (2019)
6. Smith, R., Meng, K., Dong, Z., Simpson, R.: Demand response: a strategy to address residential air-conditioning peak load in Australia. Clean Energy. **1**(3), 223–230 (2013)
7. Narayanan, R., Sethuvenkatraman, S., Pippia, R.: Energy and comfort evaluation of a fresh air-based hybrid cooling Systems in hot and Humid Climates. Energies. **15**, 7537 (2022)
8. Ausgrid: Ausgrid demand management cool saver interim report. Ausgrid, Sydney (2015)
9. Lee, S., Whaley, D.M., Saman, W.Y.: Electricity demand profile of heating and cooling appliances in Australian low energy residential buildings. Int. J. Ind. Electron. Electr. Eng. **6**(6), 33–43 (2018)
10. Narayanan, R., Parthkumar, P., Pippia, R.: Solar energy utilisation in Australian homes: a case study. Case Stud. Therm. Eng. **28**, 101603 (2021)
11. Fischer, D., Madani, H.: On heat pumps in smart grids: a review. Renew. Sust. Energ. Rev. **70**, 342–357 (2017)
12. Kok, K., Roossien, B., MacDougall, P., van Pruissen, O., Venekamp, G., Venekamp, G., Kamphuis, R., Laarakkers, J., Warmer, C.: Dynamic pricing by scalable energy management systems — field experiences and simulation results using PowerMatcher. In: 2012 IEEE Power and Energy Society General Meeting, pp. 1–8. Institute of Electrical and Electronics Engineers, San Diego (2012)
13. Le, K.X., Huang, M.J., Wilson, C., Shah, N.N., Hewitt, N.J.: Tariff-based load shifting for domestic cascade heat pump with enhanced system energy efficiency and reduced wind power curtailment. Appl. Energy. **257**, 113976 (2020)
14. Pean, T.: Heat Pump Controls to Exploit the Energy Flexibility of Building Thermal Loads, 1st edn. Springer, Cham (2021)
15. Li, Y., Mojiri, A., Rosengarten, G., Stanley, C.: Residential demand-side management using integrated solar-powered heat pump and thermal storage. Energy Build. **250**, 111234 (2021)
16. Simko, T., Luther, M.B., Li, H.X., Horan, P.: Applying solar PV to heat pump and storage Technologies in Australian Houses. Energies. **14**(17), 5480 (2021)
17. Baniasadi, A., Habibi, D., Bass, O., Masoum, M.A.S.: Optimal real-time residential thermal energy Management for Peak-Load Shifting with Experimental Verification. IEEE Trans. Smart Grid. **10**(5), 5587–5599 (2019)

18. CSIRO.: Typical house energy Use study webpage, https://ahd.csiro.au/other-data/typical-house-energy-use/. Last accessed 17 Sept 2022
19. Mitsubishi MSZ-AP DataBook.: https://www.mitsubishi-les.info/database/servicemanual/files/DataBook_2018_M_Pseries.pdf. Last accessed 17 Sept 2022
20. Commonwealth of Australia.: Water heating data collection and analysis, residential End use monitoring program (REMP) April 2012. Canberra (2012)
21. Fischer, D., Boskov Lindberg, K., Madani, H., Witter, C.: Impact of PV and variable prices on optimal system sizing for heat pumps and thermal storage. Energ. Build. **128**, 723–744 (2016)
22. AEMO.: Aggregated historical data webpage, https://aemo.com.au/energy-systems/electricity/national-electricity-market-nem/data-nem/aggregated-data. Last accessed 20 Sept 2022

Chapter 15
Comparison of Decomposition Techniques in Forecasting the Quarterly Numbers of Pole-Mounted Transformer Failures

Nhlanhla Mbuli and Jan-Harm C. Pretorius

15.1 Introduction

According to IEC 60076-1 [1], a transformer may be defined as "a static piece of apparatus with two or more windings which, by electromagnetic induction, transforms a system of alternating voltage and current into another system of voltage and current usually of different values and at the same frequency for the purpose of transmitting electrical power."

In the case of pole-mounted transformers [2], they form a vital link between the electric utility and a significant number of customers. Because of the environment in which these transformers exist, they are exposed to a variety of factors [3, 4], such as weather and climatic factors (such as lightning), environmental factors (such as trees), and human activities (such as vehicle accidents). These factors can be root causes of stresses that may culminate in the total failure of pole-mounted transformers, through [5] electrical, mechanical, and thermal modes of failure.

Costs of failures of pole-mounted transformers are related to [6] loss of revenue and the cost of repair/replacement incurred by the utility. These costs are closely linked to the frequency of failures [7, 8] and duration [9, 10] of forced outages following the failure. The duration of a forced outage [11] accounts for the time it takes to restore supply to customers, and the cost of an outage generally increases

N. Mbuli (✉)
Postgraduate School of Engineering Management, University of Johannesburg and Eskom Holdings SoC Limited, Johannesburg, South Africa
e-mail: mbulin@eskom.co.za

J.-H. C. Pretorius
Postgraduate School of Engineering Management, University of Johannesburg, Johannesburg, South Africa
e-mail: jhcpretorius@uj.ac.za

X. Wang (ed.), *Future Energy*, Green Energy and Technology,
https://doi.org/10.1007/978-3-031-33906-6_15

with its duration. Thus, immediate availability of spares is critical to minimising the duration of outages and the associated costs thereof.

From the foregoing discussion, it is apparent that in order to ensure that there is adequate amount of spares available, spares planning must be carefully undertaken a priori. Otherwise, there will [12] either be excessive amount of spares, leading to wasted opportunity costs of business money, or there will be a shortage, culminating in longer durations of pole-mounted transformer outages, leading to worsened reliability of supply.

The data on the historical number of failures of pole-mounted transformers can be captured in a time series, i.e. an ordered sequence of values that are recorded over equal intervals of time [13]. One class of models that can be used to forecast time series data comprises decomposition forecasting techniques [14]. A literature review was conducted on the applications of these techniques in power systems [15]. One of the key findings was that these methods have been utilised primarily for forecasting load, energy and power of renewable energy sources, and electricity price. No applications of these techniques in the forecasting of number of failures or spares were found.

The aim of the paper is to present a novel application of decomposition methods, i.e. multiplicative and additive, in forecasting of the quarterly number of pole-mounted transformer failures. The specific objectives and contributions of the paper are as follows:

- To fit the additive and multiplicative decomposition models to the time series data of the quarterly numbers of failed pole-mounted transformers.
- To assess the fitted models.
- To determine the accuracies of the fitted models.
- To compare the performances of the fitted models based on the calculated values of the accuracy measures.

The rest of the paper is made up of the following sections: Section 15.2 discusses the formulation of decomposition forecasting methods and thereafter the measures of expressing the accuracy of forecasting models. The methodology of the case study is discussed in Sect. 15.3. The results of the forecasted quarterly number of failures are discussed in Sect. 15.4, with the conclusions of the study drawn in Sect. 15.5.

15.2 Decomposition Forecasting Models and Measures of Accuracy [16]

This section commences by discussing the procedures for carrying out additive and multiplicative decomposition forecasting, giving the necessary steps and formulations needed. Thereafter, the measures for expressing the accuracy of the forecasting models are described, again with the associated formulations provided.

15.2.1 Multiplicative Decomposition Model

The multiplicative decomposition can be written as

$$y(t) = TR(t) \times SN(t) \times CL(t) \times IR(t) \tag{15.1}$$

where *TR*(t) is the trend, *SN* (t) is the seasonal factor, *CL*(t) is the cycle factor, and *IR*(t) is the irregular component. In multiplicative decomposition, these factors are multiplicative. The procedure for developing a time series forecast using multiplicative decomposition method may be summarised in the following steps:

1. Obtain estimates for *TR*(t) × *CL*(t).

   (a) Since in this study quarterly data will be assessed, a four-period moving average *MA* (t)will be derived.

   (b) Since, for example, the first and second moving averages correspond to second and third period and third and fourth period, respectively, it is necessary to calculate centred moving averages in order to obtain averages that correspond to the original time periods. This is achieved by using Eq. (15.2) the average of two centred moving averages, with successive centred moving averages calculated in the same manner. If an odd number of values was used in calculating the moving average, there would not have been a need for the centring procedure.

$$\text{CMA}(t) = \frac{(MA(t) + MA\,(t+1))}{2} \tag{15.2}$$

   (c) The calculated centred moving average in time period t, *CMA*(t) is considered to be equal to *TR*(t) × *CL*(t) since the averaging procedure is assumed to remove seasonal variations and any short-term irregular fluctuations, leaving the trend effect (longer-term effect) and cyclical effects intact.

2. Obtain estimates for *SN*(t) × *IR*(t).

$$SN(t) \times IR(t) = \frac{y\,(t)}{[tr\,(t) \times cl\,(t)]} \cong \frac{y\,(t)}{CMA\,(t)} \tag{15.3}$$

3. Obtain estimates for *SN*(t).Group the calculated, consecutive values of *sn*(t) × *ir*(t) into consecutive groups of four values. Calculate the value of $\widehat{sn(t)}$, i.e. the average of the values, one for each quarter, using

$$\widehat{sn(t)} \cong sn(t) \times ir(t) \tag{15.4}$$

The sum of the values of $\widehat{sn(t)}$may not add to 4, and they meet to be scaled appropriately to ensure they sum to 4. The averaging procedure in this step is considered to eliminate any short-term irregular fluctuations, while the seasonal effect remains.

4. Calculate the value of the deseasonalised observation in time period t, using Eq. (15.5), so that a better estimate of the trend can then be determined.

$$d\ (t) = \frac{y\ (t)}{sn(t)} \tag{15.5}$$

5. An appropriate polynomial trend of the type in Eq. (15.6) is fitted deseasonalised data.

$$d\ (t) = b_0 + b_1{}^1 + \cdots + b_n t^n + e(t) \tag{15.6}$$

6. At the end of step 5, the estimates of $SN(t)$ and $TR(t)$ have been determined. From Eq. (15.1), it can be shown that

$$cl(t) \times ir(t) = \frac{y\ (t)}{[tr\ (t) + sn(t)]} \tag{15.7}$$

and thus, the estimate for $CL(t) \times IR(t)$ can be obtained.

7. By calculating the three-point moving average of terms in Eq. (15.7), utilising Eq. (15.8), the values of $cl(t)$, the estimate of the cycle component can be obtained, due to the averaging process.

$$cl(t) = \frac{[cl\ (t-1) \times ir\ (t-1)] + [cl(t) \times ir(t)] + [\ cl\ (t+1) \times ir\ (t+1)]}{3} \tag{15.8}$$

8. Finally, the estimate of $ir(t)$ is calculated using Eq. (15.9).

$$ir(t) = cl(t) \times \frac{ir\ (t)}{cl\ (t)} \tag{15.9}$$

The estimates $tr(t)$, $sn(t)$, $cl\ (t)$ and $ir\ (t)$, as explained above, fully described above the time series and can be used to forecast future values of the time series. If no pattern exists in the irregular component, the value of $IR(t)$ is predicted to be one and the point forecast will be

$$\widehat{y\ (t)} = tr\ (t) \times sn(t) \times cl(t) \tag{15.10}$$

and if a well-defined cycle does not exist and cannot be predicted, *CL*(t) is taken to be 1, the point forecast becomes

$$\widehat{y\ (t)} = tr\ (t) \times sn(t) \tag{15.11}$$

15.2.2 Additive Decomposition Model

The additive decomposition model can be written as

$$y(t) = TR(t) + SN(t) + CL(t) + IR(t) \tag{15.12}$$

where, just like in the multiplicative decomposition, *TR* (t) is the trend, *SN* (t)is the seasonal factor, *CL*(t) is the cycle factor and *IR*(t) is the irregular component. In this case, the terms are additive, rather than multiplicative as is the case in multiplicative decomposition. The following steps summarise a procedure for developing a time series forecast using additive decomposition.

1. Obtain estimates for *TR*(t) + *CL*(t). The same procedure detailed in Step 1, in Sect. 15.2.1, is followed here.
2. Obtain estimates for *SN*(t) + *IR*(t). The model, as given in Eq. (15.12), implies that.

$$SN(t) + IR(t) = y(t) - [TR(t) + CL(t)] \tag{15.13}$$

and it follows that the estimate of *SN*(t) + *IR*(t), i.e. *sn*(t) + *ir*(t),is

$$sn(t) + ir(t) = y(t) - [tr(t) + cl(t)] = y(t) - MCA(t) \tag{15.14}$$

3. Obtain estimates for *SN*(t). Group the calculated, consecutive values of $sn\ (t) + ir\ (t)$ into consecutive groups of four values. Calculate the value of $\widehat{sn\ (t)}$, i.e. the average of the values, one for each quarter, using.

$$\widehat{sn\ (t)} \cong sn\ (t) + ir\ (t) \tag{15.15}$$

The sum of the values of $\widehat{sn\ (t)}$) may not add to 4, and they need to be scaled appropriately to ensure they sum to 4. The averaging procedure in this step is considered to eliminate any short-term irregular fluctuations, while the seasonal effect remains.

4. Calculate the value of the deseasonalised observation in time period t using, using Eq. (15.16), so that a better estimate of the trend can then be determined.

$$d(t) = y\ (t) - sn(t) \tag{15.16}$$

5. Fit an appropriate polynomial trend of the type in Eq. (15.6) to the deseasonalised data.
6. At the end of step 5, the estimates of $sn\ (t)$ and $tr(t)$ have been determined. From Eq. (15.12), it can be shown that

$$CL\ (t) + IR\ (t) = y\ (t) - [TR(t) + SN\ (t)] \tag{15.17}$$

and thus, the estimate for $CL\ (t) + IR\ (t)$ can be obtained from

$$cl\ (t) + ir\ (t) = y\ (t) - [tr\ (t) \times sn\ (t)] \tag{15.18}$$

7. By calculating the three-point moving average of terms in Eq. (15.18), utilising Eq. (15.8), the values of $cl\ (t)$, the estimate of the cycle component can be obtained, due to the averaging process.
8. Finally, the estimates of $ir\ (t)$ are calculated using Eq. (15.19).

$$ir\ (t) = (cl\ (t) + ir\ (t)) - cl(t) \tag{15.19}$$

The estimates $tr(t)$, $sn(t)$, $cl\ (t)$ and $ir\ (t)$, as explained above, describe the time series fully and can be utilised to forecast future values of the time series. If the irregular component demonstrates no particular pattern, the value of $IR\ (t)$ is predicted to be zero (0), and the point forecast will be

$$\widehat{y\ (t)} = tr\ (t) + sn\ (t) + cl\ (t)) \tag{15.20}$$

and if no well-defined cycle exists or cannot be predicted, $cl(t)$ is assumed to be zero, and the point forecast becomes

$$\widehat{y\ (t)} = tr\ (t) + sn\ (t) \tag{15.21}$$

15.2.3 Error of the Forecasting Model

Accuracy of the model. Several measures can be used to express the error of the forecast. Among these measures are the mean absolute deviation (MAD), given by Eq. (15.22); mean squared error (MSE), in Eq. (15.23); root mean square error (RMSE), in Eq. (15.24); and mean absolute percentage error (MAPE), in Eq. (15.25). In this paper, all the four measures of the forecast error are used.

$$\mathrm{MAD} = \frac{\sum_{i=1}^{n} |e(t)|}{n} = \frac{\sum_{i=1}^{n} \left|y(t) - \widehat{y(t)}\right|}{n} \tag{15.22}$$

$$\mathrm{MSE} = \frac{\sum_{i=1}^{n} (e(t))^2}{n} = \frac{\sum_{i=1}^{n} \left(y(t) - \widehat{y(t)}\right)^2}{n} \tag{15.23}$$

$$\mathrm{RMSE} = \sqrt{\frac{\sum_{i=1}^{n} (e(t))^2}{n}} = \sqrt{\frac{\sum_{i=1}^{n} \left(y(t) - \widehat{y(t)}\right)^2}{n}} \tag{15.24}$$

$$\mathrm{MAPE} = \sum_{i=1}^{n} \left(\left(\frac{|e(t)|}{y(t)} \right) \Big/ n \right) * (100) = \sum_{i=1}^{n} \left(\left(\frac{\left|y(t) - \widehat{y(t)}\right|}{y(t)} \right) \Big/ n \right) * (100) \tag{15.25}$$

15.3 Methodology of the Study

The data on the monthly numbers of failed pole-mounted transformers was obtained for the Eskom Gauteng Operating Unit (a region) for the period September 2005 to September 2017. The data was processed, and data grouped into quarterly numbers of failures. The key steps followed in the study were as follows:

- The data was split into in-sample portion (used to fit the forecasting models) and out-of-sample data (used to test the accuracy of the fitted models). All the quarters are used in fitting the models and values for quarters 29–36 are used to test the accuracy of the models in forecasting future values.
- Additive and multiplicative models were fitted to data. Time series is decomposed into various components from which the model is built.
- Forecasts are made and accuracy measures, using forecasted values and out-of-sample data, are determined and used to compare the utilities of the forecasting models.

15.4 Results and Discussion

The results of developing the additive and multiplicative decomposition forecasts are presented in this section. The time series of observed and forecasted values are shown in Figs. 15.1a and 15.2a, with the dotted vertical lines in these two figures, and in Figs. 15.1f and 15.2f, showing the border between the fitted model values, on the left of the line, and forecasted values on the right. The results are discussed in this section.

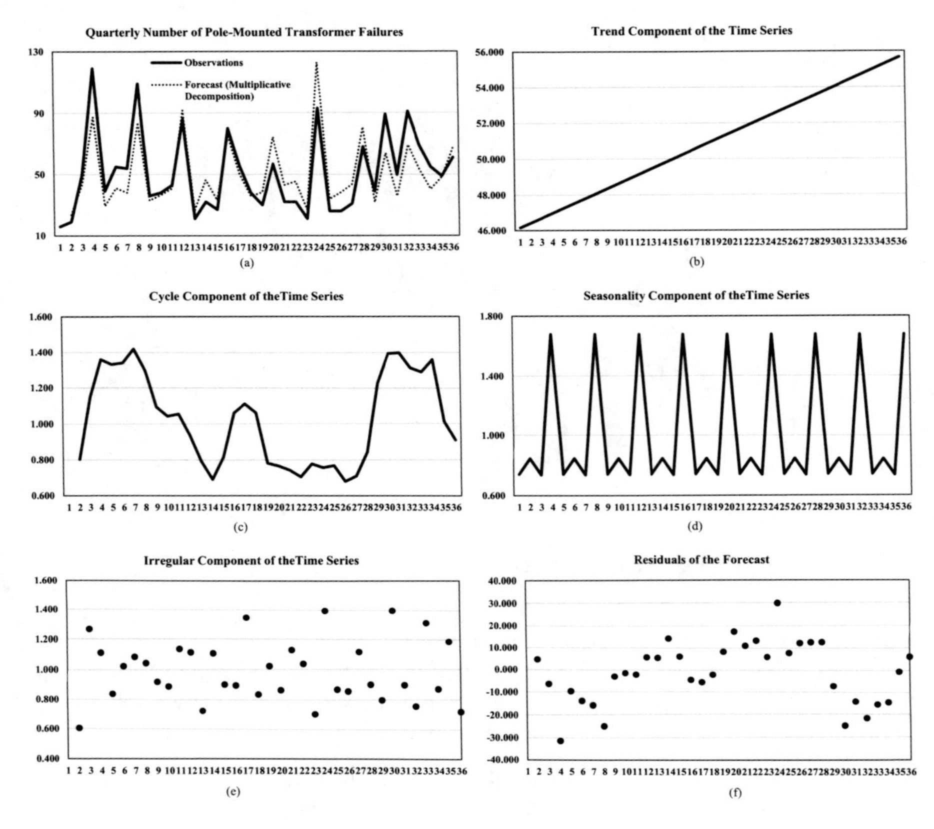

Fig. 15.1 Results of the multiplicative decomposition model construction, including (**a**) observations and forecasted values, (**b**) trend components, (**c**) seasonality, (**d**) irregular component, (**e**) cycle component, and (**f**) residuals of the forecast

15.4.1 Multiplicative Decomposition

The multiplicative trend, seasonality and irregular components of the time series of quarterly numbers of failures are shown in Fig. 15.1b–d, respectively. The values of the irregular component seem to be spread about the "*y* equals to one" line, and thus the irregular component is predicted to be equal to one in the forecasting model.

On the other hand, the cycle shows a very significant variation in its magnitudes, from just above 0.7 to around below 1.4. Thus, it is included in the forecasting model. The values of the calculated residuals, shown in Fig. 15.1f, are significantly large in relation to the magnitudes of observations, and this could reduce the calculated accuracies of the forecasting models.

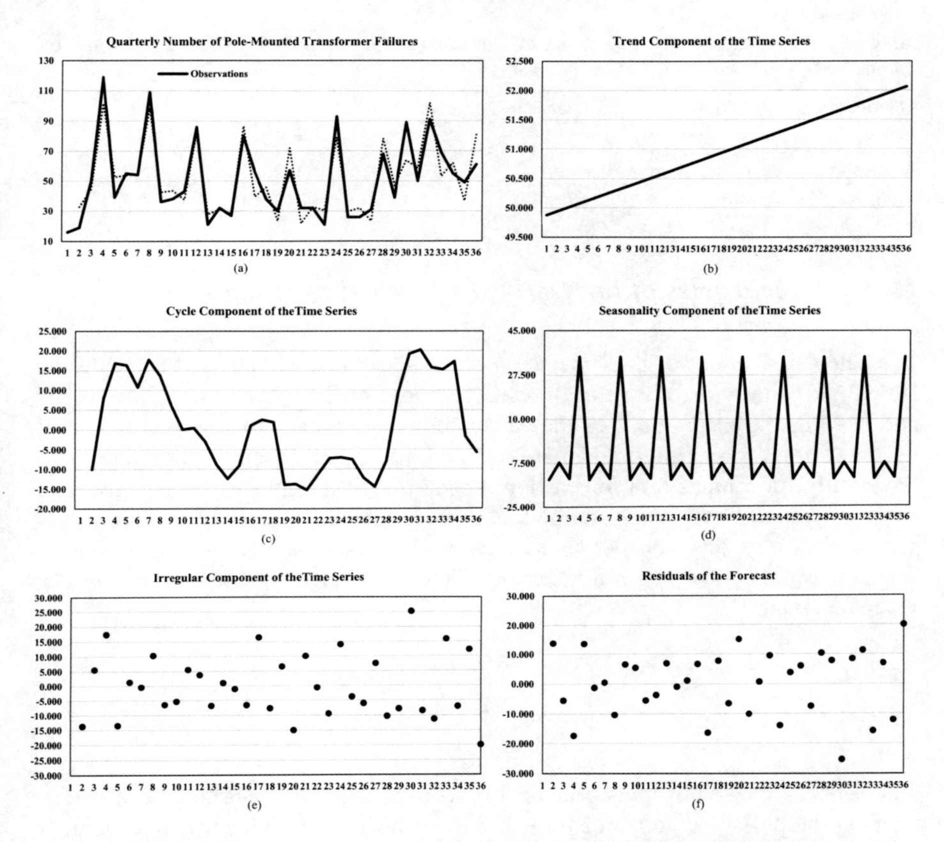

Fig. 15.2 Results of the additive decomposition model construction, including (**a**) observations and forecasted values, (**b**) the trend, (**c**) seasonality component, (**d**) irregular component, (**e**) the cycle, and (**f**) residuals of the forecast

15.4.2 Additive Decomposition

The results of additively decomposing time series of data of quarterly failures are summarised in Fig. 15.2. Following decomposition of the time series, the resulting trend, seasonality and irregular components are shown in Fig. 15.2b–d, respectively. The values of the irregular components are spread about the "*y* equals to zero line" and can be predicted to be zero, and thus the irregular component is excluded in calculating the forecasted values.

The cycle component can vary from the smallest value of -15 to a highest value of 20 and thus needs to be incorporated into the model to forecast the number of failures. The residuals, shown in Fig. 15.2f, seem to have magnitudes that are appreciably significant in relation to the values of the observations, and this could have mean that the accuracy of the forecasting model may be adversely affected.

Table 15.1 Calculated forecasting errors for the additive and multiplicative decomposition forecast methods, using various measures of error

Decomposition methods	Measures of error			
	MAD	MAPE	MSE	RMSE
Multiplicative decomposition	13.160	19.95	231.677	15.221
Additive decomposition	13.492	21.407	218.470	14.781

15.4.3 Accuracies of the Derived Forecasting Models

The calculated values of the accuracy measures, namely, MAD, MAPE, MSE and RMSE, for the two forecasting models are presented in Table 15.1. For MAD and MAPE, the magnitudes of the calculated error smaller for the multiplicative decomposition forecasting model are less and, thus, better than those obtained for the additive decomposition forecasting model. For MSE and RMSE, the values of error are smaller and, therefore, better for the additive decomposition forecasting model. Based on the four measures of error, it can be concluded that the methods compete evenly. Also, the differences in the calculated values of the measures of error are small.

15.5 Conclusion

In this paper, the authors presented the results of a study on forecasting the quarterly number of pole-mounted transformer failures using multiplicative and additive decomposition techniques. It is found that the magnitudes of the forecast residuals are appreciably high, in relation to observations, and this can have adverse impact on the accuracy of the developed models.

Secondly, there seems to be equal capabilities in these models to forecast the number of failures because, firstly, the differences in the calculated values of measures of error are small. Also, the superiority of one method over another depends on the measure of error used, and one method could be better based on selected measure of error. Future research work will assess the suitability of exponential smoothing as potential forecasting in forecasting the quarterly number of failed pole-mounted transformers.

Acknowledgement The authors are grateful for the resources provided by the University of Johannesburg and Eskom Holdings SoC towards the completion of this project.

References

1. International Standard, IEC 60076-1, Edition 2.1, Edition 2:1993 consolidated with amendment 1:1999, Power transformers –Part 1: General, 2000-04
2. Eduful, G., Mensah, G.: An investigation into protection integrity of distribution transformers – a case study, Proceedings of the World Congress on Engineering 2010, Vol. II, WCE 2010, June 30 – July 2, 2010, London, U.K., 1–6 (2010)
3. Chow, M.Y., Taylor, L.S.: A novel approach for distribution fault analysis. IEEE Trans. Power Deliv. **8**, 1882–1889 (1993)
4. Zhu, D.: Electric distribution reliability analysis considering time-varying load, weather conditions and reconfiguration with distributed generation. Ph.D. dissertation, Faculty of the Virginia Polytechnic Institute, State Univ., Blacksburg, Virginia (27 Mar. 2007)
5. Christina, A.J., Salam, M.A., Rahman, Q.M., Wen, F., Ang, S.P.: Causes of transformer failures and diagnostic methods – A review. Renew. Sustain. Energy Rev. **82**, 1442–1456 (2018)
6. Mirzai, M., Gholami, A., Aminifar, F.: Failures analysis and reliability calculation for power transformers. J. Electr. Syst. **2**, 1–12 (2006)
7. Mbuli, N., Mendu, B., Seitshiro, M., Pretorius, J.H.C.: What outage attributes affect the frequencies of forced transformer outages in Eskom regions? Evidence from 2007–2016 data. Energy Rep. **6**, 474–482 (2020)
8. Mbuli, N., Sikhakhane, S., Seitshiro, M., Pretorius, J.H.C.: Frequency of forced outages for subtransmission circuit breakers: statistical analysis of data for the period 2005–15. Przeglad Elektrotechniczny. **03**, 98–106 (2020)
9. Mbuli, N., Mendu, B., Seitshiro, M., Pretorius, J.H.C.: Statistical analysis of forced outage durations of Eskom sub transmission transformers. Energy Rep. **6**, 321–329 (2020)
10. Mbuli, N., Mendu, B., Pretorius, J.H.C.: Statistical analysis of forced outage duration data for subtransmission circuit breakers. Energy Rep. **8**, 1424–1433 (2022)
11. Shuai, M., Chengzhi, W., Shiwena, Y., Hao, G., Jufang, Y., Hui, H.: Review on economic loss assessment of power outages. Procedia Comput. Sci. **130**, 1158–1163 (2018)
12. da Silva, A.M.L., de Carvalho, C.J.G., Chowdhury, A.A.: Probabilistic methodologies for determining the optimal number of substation spare transformers. IEEE Trans. Power Syst. **25**(1), 68–77 (2010)
13. Dantas, T.M., Oliveira, F.L.C., Repolho, H.M.V.: Air transportation demand forecast through Bagging Holt Winters methods. J. Air Transp. Manag. **59**, 116–123 (2017)
14. Dragan, D., Kramberger, T., Intihar, M.: A comparison of methods for forecasting the container throughput in North Adriatic Ports, IAME 2014 Conference, Norfolk, USA, 1–20. (2014)
15. Mbuli, N., Mathonsi, M., Seitshiro, M., Pretorius, J.H.C.: Decomposition forecasting methods: a review of applications in power systems. Energy Rep. **6**, 298–306 (2020)
16. Bowerman B., Connell R.O'., Koehler A.: Forecasting, Time Series, and Regression: An Applied Approach, 4th Thomson Brooks/Cole CENGAGE Learning, USA, (2005).

Part IV
Energy-Saving Technology and Thermal Management

Chapter 16
Research on Energy-Saving Optimization of Variable Flow Air Conditioning Chilled Water System

Linqing Bao, Nan Li, and Han Qin

16.1 Introduction

The biggest advantage of the primary pump variable flow system is to realize the variable flow regulation of the chiller, which can minimize the energy consumption of the pump [1]. However, whether the primary pump variable flow system is energy-saving depends on the influence of the chilled water flow change on the efficiency of the chiller and the pump. By analyzing the characteristics of complex air conditioning systems, Zhenjun Ma et al. [2] developed the pressure drop model of different water networks in complex air conditioning systems. They used it to formulate the optimal sequential control strategy. The results show that these optimal control strategies can save about 12~32% of pump energy consumption [3]. Xing Fang et al. [4] proposed an evaluation method based on exergy analysis to evaluate the performance of central air conditioning systems. A chilled water system of central air conditioning in Changsha, China, is the research object. Based on the exergy analysis, the optimal performance of the chilled water system and equipment under specific load conditions is obtained to maximize the exergy efficiency. According to the frequency distribution of the annual load, the annual operation matching degree of the system is obtained to evaluate the design of the long-term operation chilled water system. Kuei-Peng Lee et al. [5] constructed a model by using EnergyPlus simulation software and used a hybrid optimization algorithm combining the particle swarm optimization algorithm and Hooke-Jeeves algorithm to determine the optimal setting of the chilled water system and minimize the energy consumption of the chilled water system. The results show that compared with the

L. Bao · N. Li (✉)
Chongqing University, Chongqing, China

H. Qin
China Southwest Architectural Design and Research Institute Corp., Ltd., Chengdu, China

X. Wang (ed.), *Future Energy*, Green Energy and Technology,
https://doi.org/10.1007/978-3-031-33906-6_16

traditional setting, the optimized setting reduces the total energy consumption of the chilled water system by 9.4% in summer and 11.1% in winter. Liu et al. [6–8] established the calculation model of air conditioning water systems and pointed out that under the premise of ensuring comfort requirements, the energy-saving effect of variable temperature difference control is better than that of fixed temperature difference control. Zhao et al. [9] gave the definition and online identification method of the most unfavorable branch of the air conditioning water system, proposed a control strategy based on the variable differential pressure set point of the most unfavorable branch, and tested it in a factory.

In summary, most of the literature on the study of the pipe network characteristics of the primary pump variable flow system focuses on the interaction between the ends or the overall heat transfer characteristics of the user side or stays in the qualitative analysis stage of the pipe network impedance under partial load [10]. However, there is a lack of in-depth research on the variation of system flow and impedance under different control strategies and different end control forms. There is a lack of joint research combining the overall pipe network characteristics of the primary pump variable flow system with the operation of chillers and pumps.

In this paper, the heat transfer and hydraulic model of the whole pipe network of different forms of primary pump variable flow water system are established. Then the applicable energy consumption model of chiller and variable frequency pump of different forms of primary pump variable flow system is established, respectively. The experimental platform of variable flow central air conditioning system is built. The experimental research on two kinds of water system forms with full on-off control and the combination of on-off control and continuous regulation at the end of the water system is carried out under the two control strategies of constant main pipe pressure difference and constant temperature difference, respectively. The operation characteristics of the chiller and the variable frequency pump are explored and verified with the theoretical analysis.

16.2 Establishment and Analysis of Energy Consumption Model

16.2.1 Energy Consumption Model of the Chiller

The flow-cooling thermodynamic characteristics of chilled water are different under different control strategies and different end forms. The method of measuring the effect of different forms on the energy consumption of the chiller is as follows: In the case of the same load rate and water supply temperature and no bypass water, the return water temperature and total flow rate of the main pipe are compared [11].

Fixed Trunk Pressure Difference Control (FPD) For the terminal on-off regulation control, the decisive factors of the actual flow-cooling thermal characteristic curve are the hydraulic coupling degree of the system and the random distribution of the water valve on-off state in space.

The resistance coefficient of each nonterminal branch in the original system is reduced to a minimum value (close to 0), and the resistance coefficient of the terminal branch of the original system is kept unchanged. The total resistance coefficient of this state is $S_{sys}{}'$. If the resistance coefficient of a single terminal branch is S_{ter} and there are n terminals in the system, then:

$$S'_{sys} = S_{ter}/n^2 \tag{16.1}$$

The system hydraulic coupling degree α is defined as:

$$\alpha = 1 - S_{sys}{}'/S_{sys} \tag{16.2}$$

where S_{sys} is the total resistance coefficient of the system and α is the hydraulic coupling degree of the system.

Assuming that the part of the system that does not include the root branch is the end group, and the total resistance coefficient of the end group is S_{tg}, then there is:

$$S_{tg} = S_{sys} - S_{root} \tag{16.3}$$

where S_{root} is the resistance coefficient of the root branch.

If the selection of all terminals is consistent, the flow Q_{one} of a single terminal in the system can be derived as follows:

$$Q_{one} = \sqrt{\left[\mathrm{DP}_{sys}\,(1-\alpha)\right] / \left[S_{ter}\left(k^2\alpha + 1 - \alpha\right)\right]} \tag{16.4}$$

where DP_{sys} is the pressure difference between the system water collectors, the system contains n ends, and k is the end opening rate.

In order to determine the relationship between the overall transient cooling capacity of the water system and the amount of water, it is necessary to determine the heat transfer characteristics of a single coil first [12]. The relationship between cooling capacity and flow rate of a single coil is as follows:

$$\phi_{one} = C_2 Q^{0.424} = C_2 Q_{one}{}^{0.424} \tag{16.5}$$

The total flow and cooling capacity of the system can be obtained from Eqs. 16.6 and 16.7.

$$Q_{sys} = nkQ_{one} \tag{16.6}$$

$$\phi_0 = nk\phi_{one} = nkC_2 Q_{one}{}^{0.424} \tag{16.7}$$

Therefore, the chiller input power equation is as follows:

$$P_{\text{in}} = \frac{M_r c_p}{\eta_i \eta_m \eta_d \eta_e}\left[(C_1 (t_{w2}-t_{w1})/F_R K_c + t_{w1}) - \left(t_{c\cdot w1}-kC_2{}' Q_{\text{one}}{}^{0.424}/F_{c\cdot R} K_0\right)\right] \tag{16.8}$$

where C_1 and $C_2{}'$ are proportional constants.

The end is a mixture of on-off control and continuous adjustment.

The end of this part of the on-off control is regarded as a continuous adjustment end described by Eq. 16.9. Then the total flow and total cooling capacity of the system are calculated according to the system composed of multiple continuous control valves:

$$\phi_0 = nkC_2 Q_{\text{one}}{}^{0.424} + \sum_{i=1}^{m} C_2 Q_i^{0.424} \tag{16.9}$$

The input power equation of the chiller becomes:

$$P_{\text{in}} = \frac{M_r c_p}{\eta_i \eta_m \eta_d \eta_e}\left[(C_1 (t_{w2} - t_{w1})/F_R K_c + t_{w1}) - \left(C_2{}'\phi_0/F_{c\cdot R} K_0\right)\right] \tag{16.10}$$

Fixed Temperature Difference Control (FTD) Generally, the chilled water outlet temperature of the chiller $t_{c\cdot w2}$ is controlled to be 7 °C [13]. And the temperature difference between the supply and return water of the main pipe is controlled to be 5 °C. At this time, the chilled water flow changes linearly with the load, that is:

$$\phi_0 = M_{c\cdot w} c_{c\cdot w} (t_{c\cdot w1} - t_{c\cdot w2}) = \rho Q c_{c\cdot w} \Delta t = kQ \tag{16.11}$$

Among them, Q is the volume flow of chilled water, m^3/s; ρ is the density of water, kg/m^3; Δt is the temperature difference, 5 °C; and k is a proportional constant, $k = \rho c_{c\cdot w} \Delta t$.

Therefore, the input power equation of the chiller is:

$$P_{\text{in}} = \frac{M_r c_p}{\eta_i \eta_m \eta_d \eta_e}\left[(C_1 (t_{w2} - t_{w1})/F_R K_c + t_{w1}) - \left(12 - k' Q/F_{c\cdot R} K_0\right)\right] \tag{16.12}$$

where k' is the proportional constant.

The energy efficiency ratio of the chiller is:

$$\text{EER} = \phi_0/P_{\text{in}} \tag{16.13}$$

Combining Eqs. 16.11 and 16.12 yields the following:

$$\mathrm{EER} = f_{\mathrm{EER}}\left(Q, t_{w1}, t_{w2}, \eta\right) \tag{16.14}$$

16.2.2 Energy Consumption Model of Pump

Pressure Difference Control of Stem Pipe The chilled water system is divided into I and II parts by water separator and water collector: I part is called cold-source-side, while II part is called user-side. It is assumed that the resistance coefficients of these two pipes are S_1 and S_2, respectively, and the total resistance coefficient of the system is $S = S_1 + S_2$, and then the head of the pump is:

$$H = SQ^2 = (S_1 + S_2)\, Q^2 \tag{16.15}$$

For the constant main pipe pressure difference system, the pressure difference between the water distributor and the water collector remains constant, which means $SQ^2 = C$. Therefore, Eq. 16.15 can be rewritten as Eq. 16.16, which is called the control curve. It describes the trajectory of the pump operating point during the pump frequency conversion adjustment process.

$$H = S_1 Q^2 + C \tag{16.16}$$

The output power of the pump is:

$$N_e = \gamma QH = \gamma \left(S_1 Q^3 + CQ\right) \tag{16.17}$$

The input power of the variable frequency pump can be expressed as:

$$N_{\mathrm{in}} = N_e/\eta = \left(S_1 Q^3 + CQ\right)\gamma/\eta \tag{16.18}$$

Constant Temperature Difference Control When using the temperature difference between the fixed supply and return water main pipe to control the pump frequency, the temperature difference between the supply and return water main pipe of the pipe network system always equals the design temperature difference. No matter how the end load changes, when the pump adopts the constant temperature difference control, there is a corresponding pressure difference control point in the pipe network system. When the pump frequency is controlled by the pressure difference of this point, the regulation effect of the pressure difference control method and the temperature difference control method can be consistent [10]. However, this point is dynamically moving, and the pressure difference control value is changing. The specific location and size of this point in the pipe network system are mainly related to the uneven degree of load change.

$$H = S'\, Q^2 + C' \tag{16.19}$$

The S' and C' in formula are constantly changing in the variable frequency regulation operation of the chilled water pump.

The output power of the pump is:

$$N_e = \gamma Q H = \gamma \left(S' Q^3 + C' Q \right) \quad (16.20)$$

At this time, the variable frequency pump input power equation becomes:

$$N_{\text{in}} N_{\text{in}} = N_e / \eta = \left(S' Q^3 + C' Q \right) \gamma / \eta \quad (16.21)$$

16.3 Construction of Experimental Platform

16.3.1 Overview of the Experimental Platform

The schematic diagram of the water system of the experimental platform is shown in Fig. 16.1. Room 202 has four fan coils(FP-4), room 203 has four fan coils(FP-6)and one fresh air unit, room 209 has one combined air conditioning unit, room 101 has one ceiling air conditioning unit, room 207 has two fan coils(FP-4), and room 402 has six fan coils(FP-6) and one air handle unit (AHU). Among them, the fan coil adopts on-off control through the electric two-way valve; the combined air conditioning unit adopts opening control through the continuous regulating valve.

The water system of the experimental platform can be changed into the following two forms: primary pump variable flow water system (at this time, P2, P3, and P4 do not run, and P1 directly variable frequency to achieve cold source side and the user side at the same time variable flow operation) and secondary pump variable flow water system (at this time, P3 and P4 are the secondary pump frequency operations, and P2 is the primary pump frequency operations).

Before the formal experiment, the experimental system was debugged as follows: manual test of sensor measuring points, balance test of the water system and wind system, overall system operation test, and calibration of each temperature sensor and pressure difference sensor.

16.3.2 Experimental Scheme

The experiment sets the system load rate and different system forms by opening the ends of the different numbers of rooms. The experimental conditions are listed in Table 16.1. The schemes 1–1 and 1–2 in the terminal on-off control form correspond to five opening rate conditions, respectively. The two schemes 2–1 and 2–2 in the terminal on-off control and continuous adjustment hybrid form correspond to

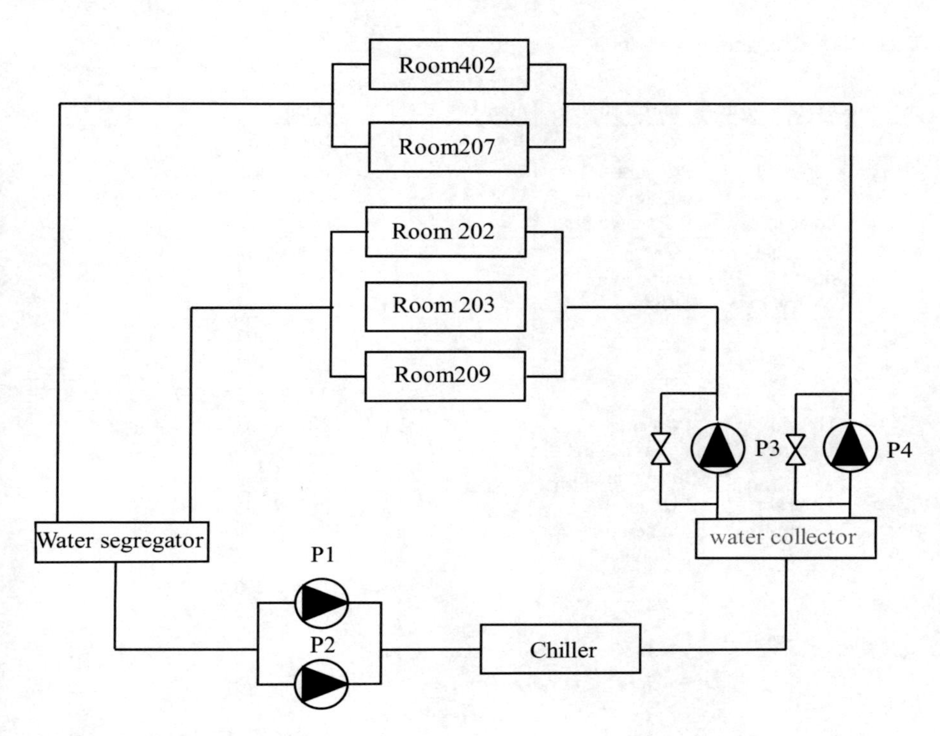

Fig. 16.1 Water system schematic

four opening rate conditions, respectively. In the experiment, the cooling water flow is kept constant, and each working condition lasts about 2 h. The outdoor meteorological parameters, indoor personnel behavior, and indoor thermal and humid environment in each working condition are stable. The sampling time of the data is 5 min.

The PPCL program is written manually in the Lunch Pad software, using control commands and calling control variables for automatic control. Instructions used by the program are mainly relational logic operation instructions, loop control instructions (PID loop control), process control instructions, and monitor point instructions.

The program of the constant temperature difference control experiment is shown in Fig. 16.2, and the program of the constant main pipe pressure difference control experiment is shown in Fig. 16.3.

Table 16.1 Experimental conditions

Scheme	System form	Control strategy	Condition number	Opening rate	Open the room
			1	100	101 + 201 + 202 + 207 + 402
1–1	Terminal on-off control	Pressure difference of fixed main pipe	2	92.5	101 + 201 + 202 + 402
1–2	Terminal on-off control	Fixed temperature difference	3	65.4	101 + 207 + 402
			4	57.9	101 + 402
			5	38.2	207 + 402
			1	100	202 + 203 + 207 + 209 + 402
2–1	Hybrid Control	Pressure difference of fixed main pipe	2	92.1	202 + 203 + 209 + 402
2–2	Hybrid Control	Fixed temperature difference	3	67.3	202 + 203 + 207 + 209
			4	59.4	202 + 203 + 209

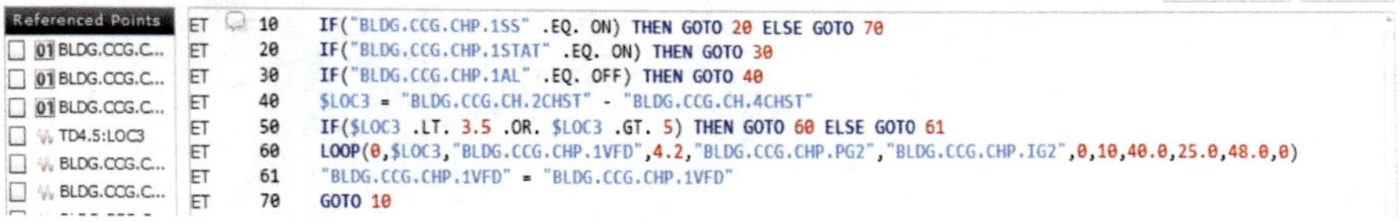

```
Referenced Points
01 BLDG.CCG.C...
01 BLDG.CCG.C...
01 BLDG.CCG.C...
TD4.5:LOC3
BLDG.CCG.C...
BLDG.CCG.C...

ET 10 IF("BLDG.CCG.CHP.1SS" .EQ. ON) THEN GOTO 20 ELSE GOTO 70
ET 20 IF("BLDG.CCG.CHP.1STAT" .EQ. ON) THEN GOTO 30
ET 30 IF("BLDG.CCG.CHP.1AL" .EQ. OFF) THEN GOTO 40
ET 40 $LOC3 = "BLDG.CCG.CH.2CHST" - "BLDG.CCG.CH.4CHST"
ET 50 IF($LOC3 .LT. 3.5 .OR. $LOC3 .GT. 5) THEN GOTO 60 ELSE GOTO 61
ET 60 LOOP(0,$LOC3,"BLDG.CCG.CHP.1VFD",4.2,"BLDG.CCG.CHP.PG2","BLDG.CCG.CHP.IG2",0,10,40.0,25.0,48.0,0)
ET 61 "BLDG.CCG.CHP.1VFD" = "BLDG.CCG.CHP.1VFD"
ET 70 GOTO 10
```

Fig. 16.2 Constant temperature difference control program

```
Referenced Points
01 BLDG.CCG.C...
01 BLDG.CCG.C...
01 BLDG.CCG.C...
BLDG.CCG.C...

ET 10 IF("BLDG.CCG.CHP.1SS" .EQ. ON) THEN GOTO 20 ELSE GOTO 51
ET 20 IF("BLDG.CCG.CHP.1STAT" .EQ. ON) THEN GOTO 30
ET 30 IF("BLDG.CCG.CHP.1AL" .EQ. OFF) THEN GOTO 40
ET 40 IF("BLDG.CCG.CH.1CHDP" .LT. 1250 .OR. "BLDG.CCG.CH.1CHDP" .GT. 1280) THEN GOTO 50 ELSE GOTO 51
ET 50 LOOP(128,"BLDG.CCG.CH.1CHDP","BLDG.CCG.CHP.1VFD","BLDG_CCG_CHP_DPSET1","BLDG.CCG.CHP.PG","BLDG.CCG.CHP.IG'
ET 51 GOTO 10
```

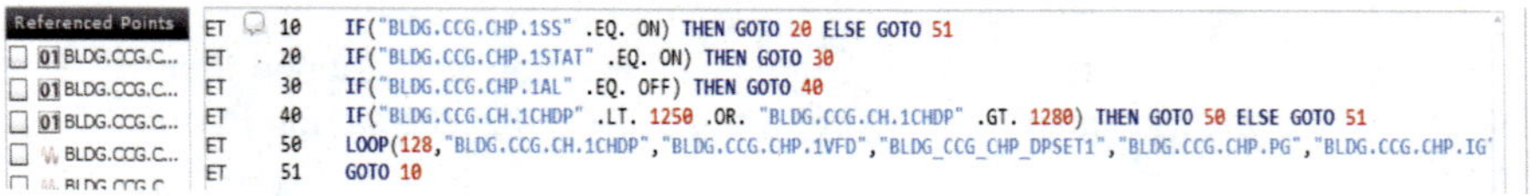

Fig. 16.3 Constant pressure difference control program

16.4 Result and Discussion

16.4.1 Comprehensive Energy Consumption Comparison of Terminal On-Off Control System

Table 16.2 shows the operating power of the chiller and pump under various operating conditions in the on-off control. It can be seen from the figure that as the system load rate decreases, the comprehensive power of the chiller and the pump decreases first and then increases. The total power is the smallest at working condition 4, which is 15.07 kW and 14.97 kW, respectively. When the load rate

Table 16.2 On-off control experiment energy consumption results

Control strategy	Condition number	Load rate (%)	Chiller/kW	Pump/kW	Summation/kW
Fixed pressure difference of main pipe	1	60.57%	18.72	1.49	20.21
	2	56.92%	17.62	1.39	19.01
	3	52.59%	18.07	0.95	19.02
	4	44.46%	14.27	0.80	15.07
	5	27.35%	16.41	0.69	17.10
Fixed temperature difference	1	67.11%	20.59	0.94	21.53
	2	58.85%	19.16	0.86	20.02
	3	51.55%	16.46	0.96	17.42
	4	45.72%	14.21	0.76	14.97
	5	24.61%	15.46	0.42	15.88

Table 16.3 Comparison of energy-saving rates between two control strategies

Condition number	1	2	3	4	5
Fixed pressure difference of main pipe	7.53%	13.00%	12.96%	31.05%	21.74%
Fixed temperature difference	1.45%	8.36%	20.27%	31.49%	27.34%

decreases, the "rebound" phenomenon occurs in condition 5, and the total power increases to 17.10 and 15.88 kW, respectively.

From Table 16.3, it can be seen that the comprehensive energy-saving rate of the two control modes is the highest and close to each other at working condition 4. The pressure difference control of the fixed main pipe is 31.05%, and the temperature difference control is 31.49%. The comprehensive energy-saving rate of constant temperature difference control is higher than that of constant temperature difference control under working conditions 1 and 2, and the load rate of constant temperature difference control is slightly higher than that of constant temperature difference control under these two working conditions. The comprehensive energy-saving rate of the constant main pipe pressure difference control is lower than that of the constant temperature difference control under condition 3 and condition 5, and the load rate of the constant temperature difference control is slightly lower than that of the constant main pipe pressure difference under these two conditions [14]. Through comprehensive analysis, although the energy-saving rate of variable frequency pump under constant temperature difference control is about 15~30% higher than that of constant main pipe pressure difference control under working conditions 1, 2, and 5, its energy-saving advantage is reduced due to its high energy consumption of chiller, and even the comprehensive energy-saving rate is lower than that of constant main pipe pressure difference control.

Table 16.4 Energy-saving results of on-off control and continuous modulation hybrid system

Control strategy	Condition number	Load rate (%)	Chiller/kW	Pump/kW	Summation/kW
Fixed pressure difference of main pipe	1	53.21%	19.86	1.30	21.16
	2	47.06%	18.40	1.06	19.46
	3	45.44%	18.74	0.94	19.68
	4	34.76%	17.96	0.68	18.64
Fixed temperature difference	1	56.28%	20.44	1.11	21.55
	2	51.44%	18.74	0.92	19.66
	3	36.42%	15.76	0.65	16.41
	4	35.78%	14.32	0.68	15.00

Table 16.5 Comparison of energy-saving rates between two control strategies

Condition number	1	2	3	4
Fixed pressure difference of main pipe	3.14%	10.94%	9.95%	14.67%
Fixed temperature difference	1.37%	10.01%	24.89%	31.34%

16.4.2 Comprehensive Energy Consumption Comparison of Hybrid System

Table 16.4 shows the operating power of the chiller and pump under various operating conditions in the hybrid control of end on-off and continuous regulation. With the decrease of load rate, the operation power of chiller and water pump decreases, but the energy consumption of chiller is dominant. Under the control strategy of constant main pipe pressure difference, the total power of the chiller and pump decreases with the decrease in load rate. From condition 1 to condition 4, the load rate decreases from 53.21% to 34.76%, while the total power decreases from 21.16 kW to 18.64 kW. Under the constant temperature difference control strategy, the total power of the chiller and the pump decreases greatly with the decrease of the load rate. From condition 1 to condition 4, the load rate decreases from 56.28% to 35.78%, and the total power decreases from 21.55 kW to 15.00 kW.

From Table 16.5, it can be seen that the comprehensive energy-saving rate of the two control modes is the highest at condition 4, the fixed main pipe pressure difference control is 14.67%, and the fixed temperature difference control is 31.34%; the comprehensive energy-saving rate of the fixed main pipe pressure difference control is higher than that of the fixed temperature difference under the condition 1 and condition 2. The load rate of the fixed temperature difference control is slightly higher than that of the fixed main pipe pressure difference under these two conditions. The comprehensive energy-saving rate of constant main pipe pressure difference control is lower than that of constant temperature difference under working conditions 3 and 4. Due to the lower condensing temperature of the chiller controlled by constant temperature difference under these two working conditions, its operating efficiency is higher.

16.5 Conclusion

In this paper, combined with the overall heat transfer characteristics of primary pump variable flow system under different control strategies and different terminal control forms, the energy consumption model of the chiller and variable frequency pump are established, respectively. By building a variable flow centralized air conditioning system test bench, the similarities and differences of the comprehensive energy consumption and comprehensive energy-saving rate of the chiller and pump under the two control strategies are compared and studied. The main research results are as follows:

Although the energy-saving rate of the variable frequency pump is large under the constant temperature difference control, the rated power of the pump in this experiment only accounts for about one-tenth of the rated power of the chiller. From the perspective of the comprehensive energy-saving rate, the energy-saving effect of the variable frequency pump is weakened, and even the comprehensive energy-saving rate of the constant temperature difference control under some working conditions is less than that of the constant main pipe pressure difference control. Therefore, when measuring the energy-saving potential of the primary pump variable flow system, only focusing on the energy-saving rate of the variable frequency pump may exaggerate the energy-saving effect of the system, and the energy-saving effect of constant temperature difference control strategy is not significantly better than that of constant main pipe pressure difference control strategy.

References

1. Ding, Y., Su, H., Liu, K., et al.: Robust commissioning strategy for existing building cooling system based on quantification of load uncertainty [J]. Energy Build., 225 (2020)
2. Ma, Z.J., Wang, S.W.: Energy efficient control of variable speed pumps in complex building central air-conditioning systems [J]. Energy Build. **41**(2), 197–205 (2009)
3. Fang, X., Jin, X., Du, Z., et al.: Evaluation of the design of chilled water system based on the optimal operation performance of equipments [J]. Appl. Therm. Eng., 113435–113448 (2017)
4. Fang, X., Jin, X.Q., Du, Z.M., et al.: Evaluation of the design of chilled water system based on the optimal operation performance of equipments [J]. Appl. Therm. Eng., 113435–113448 (2017)
5. Lee, K.-P., Cheng, T.-A.: A simulation-optimization approach for energy efficiency of chilled water system [J]. Energy Build., 54290–54296 (2012)
6. Liu, X.F., Liu, J.P., Lu, J.D., et al.: Research on operating characteristics of direct-return chilled water system controlled by variable temperature difference [J]. Energy. **40**(1), 236–249 (2012)
7. Heydari, A., Kargar, S., Asme: Theoretical modeling and optimizing design of a packaged liquid chiller [M]. Proceedings of the Asme/Jsme Thermal Engineering Summer Heat Transfer Conference 2007. **1**, 527–537 (2007)
8. Huang, T., Liang, C., Bai, X., et al.: Study on the feature-recognition-based modeling approach of chillers [J]. Int. J. Refrig-Revue Internationale Du Froid, 100326–100334 (2019)

9. Zhao, T., Ma, L., Zhang, J.: An optimal differential pressure reset strategy based on the most unfavorable thermodynamic loop on-line identification for a variable water flow air conditioning system [J]. Energy Build., 110257–110268 (2016)
10. Wei, D., Zuo, M., Yu, J.: Control strategy for energy saving of refrigerating station systems in public buildings [J]. J. Build. Eng., 44 (2021)
11. Wu, T., Zhao, H.: An energy-saving and velocity-tracking control design for the pipe isolation tool [J]. Adv. Mech. Eng. **11**(4) (2019)
12. Fang, Z., Tang, T., Su, Q., et al.: Investigation into optimal control of terminal unit of air conditioning system for reducing energy consumption [J]. Appl. Therm. Eng., 177 (2020)
13. Ma, Z., Wang, S.: Test and evaluation of energy saving potentials in a complex building central chilling system using genetic algorithm [J]. Build. Serv. Eng. Res. Technol. **32**(2), 109–126 (2011)
14. Liu, X.-F., Liu, J.-P., Lu, J.-D., et al.: Research on operating characteristics of direct-return chilled water system controlled by variable temperature difference [J]. Energy. **40**(1), 236–249 (2012)

Chapter 17
Parametric Analysis and Comparative Study of the Transcritical CO_2 Cycle and the Organic Rankine Cycle for Low-Temperature Geothermal Sources

Kun Hsien Lu, Hsiao Wei Chiang, and Pei Jen Wang

17.1 Introduction

The soaring energy demand and the fossil fuel consumption have invoked environmental concerns. Geothermal is abundant in the nature and it does not require an additional thermal storage unit due to its non-intermittent availability. Moreover, the power generation using geothermal energy produces almost no pollutants if it is properly processed. The geothermal energy is also suitable to serve as the distributed power supply of the residence nearby to relieve the reliance from the grid. Therefore, exploiting the geothermal energy is an attractive way to deal with the energy demand.

However, most of the geothermal energy exists with relatively low temperature at 100–220 °C [1], which will be thermally inefficient if the conventional steam-based power units are applied. Organic Rankine cycle (ORC) is a promising technology to exploit the low-temperature heat sources, and it has been widely adopted for geothermal applications [2]. R245fa is one of the most frequently used working fluids in ORCs [3–6]. Nevertheless, the high global warming potential (GWP) of R245fa has induced restrictions on its future applications. As such, the investigations of low GWP refrigerants for the ORC are emerging [7–9].

Recently, supercritical CO_2 (S-CO_2) cycles are attracting interests in heat-to-power applications [10–12]. The well-matched temperature profile between the S-CO_2 and the heat source results in a more efficient energy conversion process compared to the subcritical cycles [12]. In addition, CO_2 is a natural fluid with nontoxicity, non-flammability, and high compatibility with most materials. The S-CO_2 cycles that fully operated above the critical point are suitable for utilizing

K. H. Lu (✉) · H. W. Chiang · P. J. Wang
National Tsing Hua University, Hsinchu City, Taiwan
e-mail: khlu4240@gapp.nthu.edu.tw

X. Wang (ed.), *Future Energy*, Green Energy and Technology,
https://doi.org/10.1007/978-3-031-33906-6_17

Table 17.1 Fluid characteristics [17]

	R245fa	R601a	CO_2
ODP[a]	0	0	0
GWP	1050	~20	1
T_{crit} (°C)[b]	154	187.2	31.0
P_{crit} (MPa)[b]	3.65	3.38	7.38

[a]Ozone depletion potential
[b]Critical point

high-temperature sources [13], while the transcritical CO_2 (T-CO_2) cycles are more recommended for low-temperature heat sources [14].

Zare et al. [15] investigated the geothermal-driven ORCs from both thermodynamic and economic perspectives with heat source temperature at 160–170 °C. The simple layout presents the highest net power output, the lowest total cost, and the shortest payback period compared to the regenerative and the recuperative layouts. Habibollahzade et al. [16] compared T-CO_2 cycle, S-CO_2 cycle, and ORC with a 170 °C geothermal source. The results show that the T-CO_2 cycle has the highest energy performance and the lowest payback period. In addition, the authors also suggested that lower heat source temperatures are favorable for all the cycles studied. Li et al. [3] compared a R245fa-ORC and a T-CO_2 cycle using 120–260 °C of heat source temperature and 20–0 °C of cooling temperature. The results show that the energy and exergy performance of the R245fa-ORC are both higher than the T-CO_2 cycle.

In this study, the T-CO_2 cycle and the ORCs are analyzed and compared in terms of the net power output and the heat recovery efficiency with hot water at 90–150 °C ($T_{h,\,in}$) and ambient temperature at 20–10 °C (T_{amb}) to simulate possible geothermal sources, i.e., directly from the production well or within the process of the existing geothermal power units. The recuperator is not considered in this study due to its low cost-effectiveness with low-temperature heat sources or if the turbomachinery has been well designed [8, 15, 18]. R245fa and R601a are selected as the working fluid for the ORCs and the fluid characteristics are collected in Table 17.1. Although superheating is theoretically unnecessary for both R245fa and R601a since they are not wet fluids, superheating is practically adopted to ensure no droplets occur in the expansion process. So superheating is considered for the ORCs in this study.

17.2 Method

17.2.1 Cycle Description

The ORCs in this study are operated subcritically as the same as most of the existing ORCs [19]. The working principle and the state points of the ORCs and the T-CO_2 cycle are illustrated with the temperature-entropy (T-s) diagrams shown in Fig. 17.1. The cycle configuration adopted for all the fluids is a simple Rankine cycle that

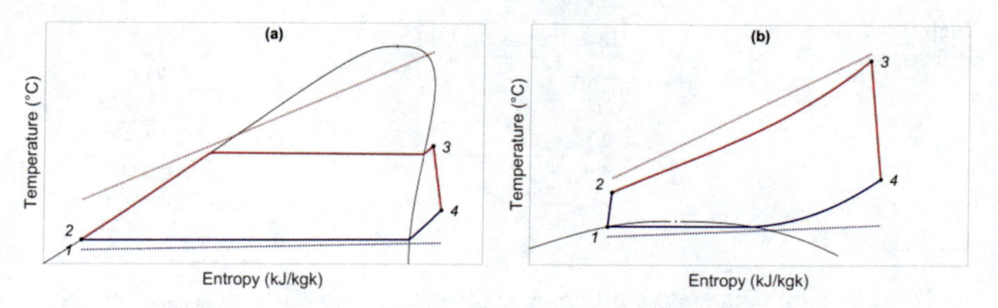

Fig. 17.1 T-s diagrams of (**a**) subcritical ORC and (**b**) T-CO2 cycle

consists of pump, evaporator (ORC)/heater (T-CO_2 cycle), expander, and condenser. Firstly, the working fluid is pressurized to the working pressure through the pump (processes 1–2). Then, the high-pressure working fluid is heated by the heat source in the evaporator/heater (processes 2–3). Afterward, the high-temperature and high-pressure energy of the working fluid can be extracted via the expander (processes 3–4). Finally, the low-pressure working fluid is condensed to liquid state and the cycle is completed.

17.2.2 Energy Analysis

The main correlations of the energy analysis for calculating the performance indicators are collected as follows:

$$Q_h = \dot{m}_{wf} (h_3 - h_2) = \dot{m}_h \left(h_{h,\text{in}} - h_{h,\text{out}}\right) \tag{17.1}$$

$$W_{\text{net}} = \dot{m}_{wf} (h_3 - h_4) - \dot{m}_{wf} (h_2 - h_1) \tag{17.2}$$

$$\eta_H = \frac{h_{h,\text{in}} - h_{h,\text{out}}}{h_{h,\text{in}} - h_0} \tag{17.3}$$

Q_h is the heat transfer rate from the heat source and h_i is the enthalpy of the fluid. $\dot{m}_i$ is the mass flow rate, where the subscript *wf* and *h* represent the working fluid and the heat source, respectively. W_{net} and η_H are the net power output and the heat recovery efficiency, respectively. The subscript 0 stands for the ambient condition.

For simplicity, this study is carried out based on the following assumptions:

- All the fluids are operated as steady-state flow.
- The working fluid is condensed to saturated liquid state before entering the pump.
- The condensing temperature is 7 K above the ambient temperature.
- Two percent pressure loss is considered in each heat exchanger.

Table 17.2 Parameter settings

Fixed parameters	Value	Variables	Value
η_p, η_p	85%	$T_{h,\text{in}}$	90/120/150 °C
T_{sh}	5 K	T_{amb}	10/15/20 °C
ΔT_{min}	5 K	P_2 (ORC)	$2{*}P_1$–$7{*}P_1$
ΔP_{loss}	2%		Or $2{*}P_1$–P_{crit}
$\dot{m}_h$	10 kg/s	P_2 (T-CO_2)	1–2 MPa

- The minimum temperature difference in the evaporator/heater (ΔT_{min}) is 5 K.
- The isentropic efficiency of pump (η_p) and expander (η_e) are 85% [3, 20, 21].

The simulation process based on the energy analysis is written in MATLAB (2019b) and the fluid properties are read from NIST REFPROP (v9.0) [22]. Firstly, the fixed parameters and the variables listed in Table 17.2 are imported into the mathematical model. Then, the working pressure (P_2) of each cycle is increased from the lower limit with a step increase of 20 kPa. Secondly, the expander inlet temperature (T_3) for the ORCs is calculated from the corresponding evaporation temperature at P_2 and 5 K superheat (T_{sh}). For the T-CO_2 cycle, T_3 is assumed constantly 5 K below $T_{h,\text{in}}$. The heat source outlet temperature ($T_{h,\text{out}}$) of each cycle is determined under the ΔT_{min} restriction during the heating process. Then, $\dot{m}_{wf}$ can be calculated by using Eq. (17.1). Finally, W_{net} and η_H can be derived from Eqs. (17.2) and (17.3), respectively. The optimal design point is selected based on the maximal W_{net}.

17.3 Results

17.3.1 Comparisons Based on the Net Power Output, W_{net}

Figure 17.2 shows the W_{net} of each cycle under various $T_{h,\text{in}}$ and T_{amb} conditions and the comparisons between each cycle are presented in Fig. 17.3. As $T_{h,\text{in}}$ increases, the W_{net} increments of the ORCs are higher than the T-CO_2 cycle and the W_{net} of each cycle becomes less sensitive to the T_{amb} variation. On the other hand, as T_{amb} rises, the W_{net} of the T-CO_2 cycle decreases more rapidly than the ORCs. Taking the case of 90 °C $T_{h,\text{in}}$ for example, 44% W_{net} decrement is found in the T-CO_2 cycle as T_{amb} increases from 10 to 20 °C, while around 30% W_{net} reduction are occurred in the ORCs. Additionally, the enhancement of W_{net} resulting from the $T_{h,\text{in}}$ increasing is more obvious with higher T_{amb}.

From Fig. 17.3, it can be clearly seen that the T-CO_2 cycle has dominant advantage in terms of W_{net} in lower-temperature environment. However, this advantage is diminished as $T_{h,\text{in}}$ and T_{amb} rise. In the case of 20 °C T_{amb} and 150 °C $T_{h,\text{in}}$, the W_{net} of the T-CO_2 cycle is the lowest and R245fa produces around 9.1% more W_{net} than the T-CO_2 cycle. For the ORCs, R245fa provides slightly higher W_{net} than R601a, especially in higher-temperature environment.

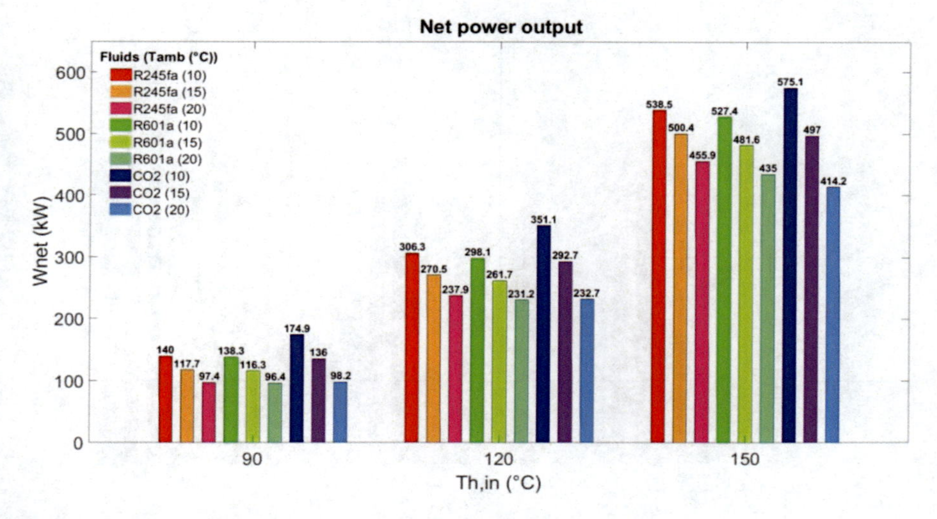

Fig. 17.2 W_{net} of each cycle under various $T_{h,\,in}$ and T_{amb}

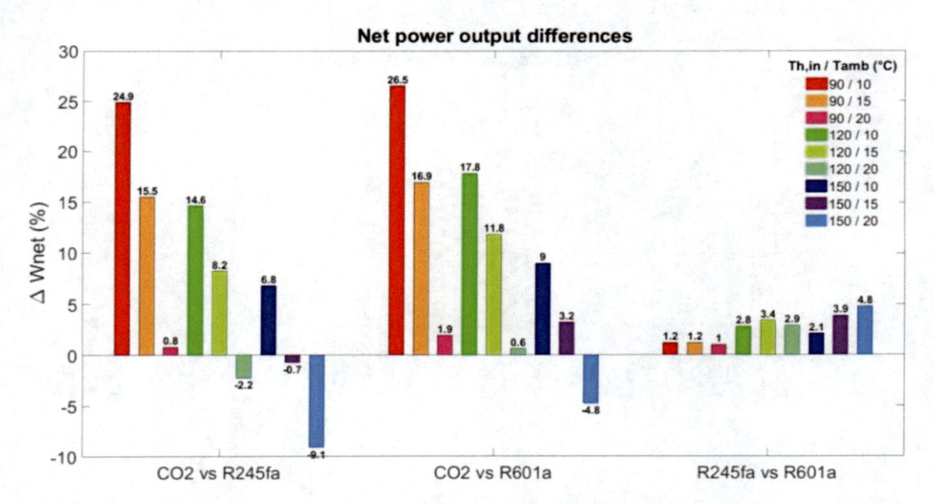

Fig. 17.3 Comparisons of W_{net} between each cycle under various $T_{h,\,in}$ and T_{amb}

17.3.2 Comparisons Based on the Heat Recovery Efficiency, η_H

The η_H of each cycle under various $T_{h,\,in}$ and T_{amb} and the comparisons between each cycle are presented in Figs. 17.4 and 17.5, respectively. From Fig. 17.4, it can be seen that the η_H of all cycles are less sensitive to $T_{h,\,in}$ and T_{amb} than W_{net}. In addition, the effect of T_{amb} on η_H is less significant than $T_{h,\,in}$, especially for the ORCs. The η_H of the T-CO_2 cycle is generally higher than the ORCs; however, the η_H of the ORCs increases more significantly with $T_{h,\,in}$. As $T_{h,\,in}$ increases from 90

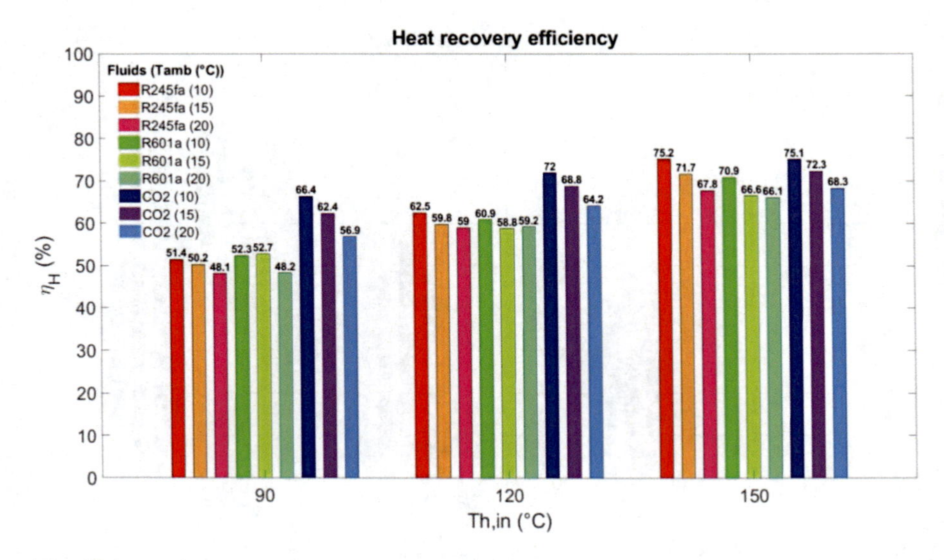

Fig. 17.4 η_H of each cycle under various $T_{h,\text{ in}}$ and T_{amb}

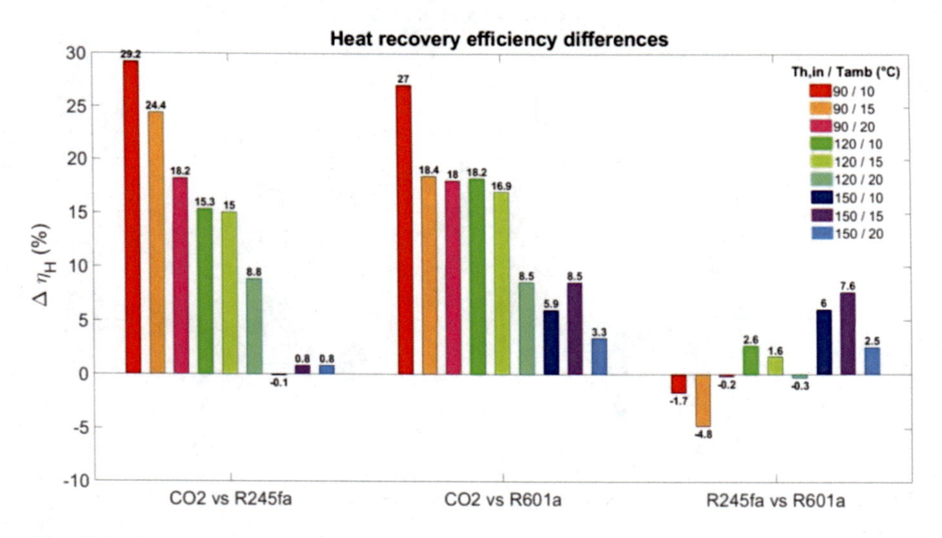

Fig. 17.5 Comparisons of η_H between each cycle under various $T_{h,\text{ in}}$ and T_{amb}

to 150 °C, the η_H of the R245fa-ORC can be increased by 40.8% to 46.12%, while 26.4% to 37.1% and only 8.3% to 12.8% increment of η_H can be obtained with the R601a-ORC and the T-CO_2 cycle, respectively.

Therefore, as can be observed from Fig. 17.5, the advantage of the T-CO_2 cycle over the ORCs in terms of η_H is more pronounced with lower $T_{h,\text{ in}}$ and T_{amb}. For the ORCs, the R601a-ORC has higher η_H than the R245fa-ORC with 90 °C of $T_{h,\text{ in}}$.

However, the R245fa-ORC performs better than the R601a-ORC in terms of η_H with higher $T_{h,\text{ in}}$ since it has greater η_H increment resulting from the $T_{h,\text{ in}}$ increasing. In fact, with 150 °C of $T_{h,\text{ in}}$, the R245fa-ORC and the T-CO_2 cycle have similar η_H that are both higher than the R601a-ORC.

17.4 Conclusions

The T-CO_2 cycle and the ORCs using R245fa and R601a are investigated in terms of W_{net} and η_H under various conditions of $T_{h,\text{ in}}$ and T_{amb}. The behaviors of each cycle are different as $T_{h,\text{ in}}$ and T_{amb} change. As $T_{h,\text{ in}}$ increases, the increments of W_{net} and η_H of the ORCs are greater than the T-CO_2 cycle. In addition, the W_{net} of all cycles becomes less sensitive to the T_{amb} with higher $T_{h,\text{ in}}$. On the other hand, as T_{amb} rises, the W_{net} of the T-CO_2 decreases more significantly than the ORCs. For each cycle, the enhancement of W_{net} resulting from the $T_{h,\text{ in}}$ increasing is more obvious with higher T_{amb}. Nevertheless, T_{amb} has less influence on η_H than $T_{h,\text{ in}}$.

In general, based on the assumptions made in this study, the T-CO_2 cycle is more recommended with lower $T_{h,\text{ in}}$ and T_{amb} since it can provide higher W_{net} and η_H than the ORCs, while the R245fa-ORC performs better as $T_{h,\text{ in}}$ and T_{amb} increase.

References

1. Hettiarachchi, H.M., et al.: Optimum design criteria for an organic Rankine cycle using low-temperature geothermal heat sources. Energy. **32**(9), 1698–1706 (2007)
2. Tartière, T., et al.: A world overview of the organic Rankine cycle market. Energy Procedia. **129**, 2–9 (2017)
3. Li, L., et al.: Thermodynamic analysis and comparison between CO2 transcritical power cycles and R245fa organic Rankine cycles for low grade heat to power energy conversion. Appl. Therm. Eng. **106**, 1290–1299 (2016)
4. Yamaguchi, T., et al.: Experimental study for the small capacity organic Rankine cycle to recover the geothermal energy in Obama hot spring resort area. Energy Procedia. **160**, 389–395 (2019)
5. Zhang, X., et al.: Economic analysis of organic Rankine cycle using R123 and R245fa as working fluids and a demonstration project report. Appl. Sci. **9**(2), 288 (2019)
6. Kong, R., et al.: Thermodynamic performance analysis of a R245fa organic Rankine cycle (ORC) with different kinds of heat sources at evaporator. Case Stud. Therm. Eng. **13**, 100385 (2019)
7. Yang, J., et al.: Simultaneous experimental comparison of low-GWP refrigerants as drop-in replacements to R245fa for Organic Rankine cycle application: R1234ze (Z), R1233zd (E), and R1336mzz (E). Energy. **173**, 721–731 (2019)
8. Song, J., et al.: Thermo-economic optimization of organic Rankine cycle (ORC) systems for geothermal power generation: a comparative study of system configurations. Front. Energy Res. **8**, 6 (2020)
9. Zhar, R., et al.: A comparative study and sensitivity analysis of different ORC configurations for waste heat recovery. Case Stud. Therm. Eng. **28**, 101608 (2021)

10. Cayer, E., et al.: Parametric study and optimization of a transcritical power cycle using a low temperature source. Appl. Energy. **87**(4), 1349–1357 (2010)
11. Kim, Y., et al.: Transcritical or supercritical CO2 cycles using both low- and high-temperature heat sources. Energy. **43**(1), 402–415 (2012)
12. Sarkar, J.: Review and future trends of supercritical CO2 Rankine cycle for low-grade heat conversion. Renew. Sustain. Energy Rev. **48**, 434–451 (2015)
13. Guo, J.-Q., et al.: A systematic review of supercritical carbon dioxide (S-CO2) power cycle for energy industries: technologies, key issues, and potential prospects. Energy Convers. Manag. **258**, 115437 (2022). https://www.sciencedirect.com/science/article/pii/S0196890422002333
14. Liu, L., et al.: Supercritical carbon dioxide (s-CO2) power cycle for waste heat recovery: a review from thermodynamic perspective. Processes. **8**(11), 1461 (2020)
15. Zare, V., et al.: A comparative exergoeconomic analysis of different ORC configurations for binary geothermal power plants. Energy Convers. **105**, 127–138 (2015)
16. Habibollahzade, A., et al.: Comparative thermoeconomic analysis of geothermal energy recovery via super/transcritical CO2 and subcritical organic Rankine cycles. Energy Convers. Manag. **251**, 115008 (2022)
17. Calm, J.M., et al.: Physical, safety, and environmental data for current and alternative refrigerants. In: Proceedings of 23rd International Congress of Refrigeration, pp. 21–26 (2011)
18. Lu, K.-H., et al.: Sensitivity analysis of transcritical CO2 cycle performance regarding isentropic efficiencies of turbomachinery for low temperature heat sources. Energies. **15**(23), 8868 (2022)
19. Astolfi, M., et al.: Selection maps for ORC and CO2 systems for low-medium temperature heat sources. Energy Procedia. **129**, 971–978 (2017)
20. Yoon, S.Y., et al.: Comparison of micro gas turbine heat recovery systems using ORC and trans-critical CO2 cycle focusing on off-design performance. Energy Procedia. **129**, 987–994 (2017)
21. Bellos, E., et al.: Investigation of a novel CO2 transcritical organic rankine cycle driven by parabolic trough solar collectors. Appl. Syst. Innov. **4**(3), 53 (2021)
22. Lemmon, E.W., et al.: NIST standard reference database 23: reference fluid thermodynamic and transport properties-REFPROP (2013)

Chapter 18
Structural Optimization of Liquid-Cooled Battery Modules with Different Flow Configurations

Kangdi Xu, Hengyun Zhang, and Jiajun Zhu

18.1 Introduction

Lithium-ion batteries have been widely used in electric vehicles because of their high energy density, long service life, and low self-discharge rate and gradually become the ideal power source for new energy vehicles [1, 2]. However, Li-ion batteries still face thermal safety issues [3, 4]. Therefore, a properly designed battery thermal management system (BTMS) is important to stabilize the battery module temperature [5, 6].

Nowadays, battery thermal management methods include air cooling [7], liquid cooling, and phase change cooling [8]. In comparison, liquid cooling has better heat transfer performance. Zhao et al. [9] arranged serpentine channels on the surface of cylindrical cells to cool their cell modules and obtained good uniform temperature performance at a 5C discharge multiplier.

In this paper, a new type of liquid-cooled shell structure is proposed. A battery module experimental platform was built according to the optimized structure, and the experimental study of thermal performance under different charge/discharge multiplier and flow rate conditions was conducted. The experiments verified that the new liquid-cooled shell with optimal inlet/outlet configuration can provide good thermal management of the battery module.

K. Xu · H. Zhang (✉) · J. Zhu
School of Mechanical and Automotive Engineering, Shanghai University of Engineering Science, Shanghai, China
e-mail: zhanghengyun@sues.edu.cn

X. Wang (ed.), *Future Energy*, Green Energy and Technology,
https://doi.org/10.1007/978-3-031-33906-6_18

18.2 Numerical Models and Experimental Studies

18.2.1 New Battery Module Liquid-Cooled Shell Model

In this paper, a new type of liquid-cooled shell structure is proposed, as shown in Fig. 18.1. The liquid-cooled shell is equipped with 4 × 5 through-holes to accommodate 18,650 Li-ion batteries, with multiple horizontal and vertical flow channels built in between the batteries.

A battery module liquid cooling experimental system was built, including a circulating thermostatic water tank, a flow meter, a charge/discharge tester, a differential pressure meter, and a temperature data acquisition system.

18.2.2 Selection of the Optimal Flow Channel Combination

This section investigates the effect of the locations and number of inlets and outlets for the liquid-cooled shell on the thermal performance of the battery module. The specific arrangement is shown in Fig. 18.2. The combination Case 1 corresponding to the maximum expected function value is selected as the optimal flow path combination. To facilitate the comprehensive consideration of the maximum temperature, temperature difference, and pressure drop on the thermal performance of the battery module, the expectation function f_D is used here for comprehensive evaluation and selection of the optimal flow channel combination configuration. The expectation function f_D is defined by the desirability functions of maximum temperature dTmax, temperature difference dT, and pressure drop dp as follows.

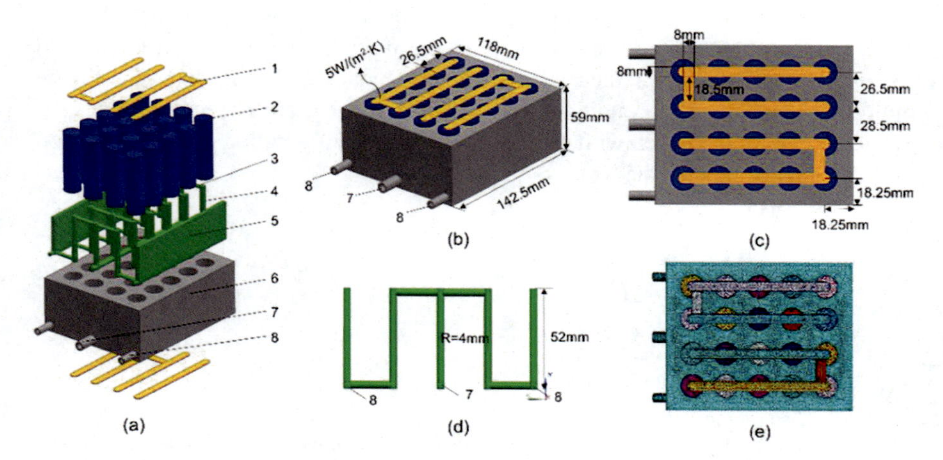

Fig. 18.1 Schematic diagram of the novel liquid-cooled shell battery module: (**a**) overall structure of battery module system; (**b**) 3D numerical model of battery module; (**c**) top view of battery module; (**d**) liquid channel structure; (**e**) grid model. 1-busbar, 2-cell, 3-lateral channel, 4-longitudinal channel, 5-liquid channel, 6-shell, 7-inlet, 8-outlet

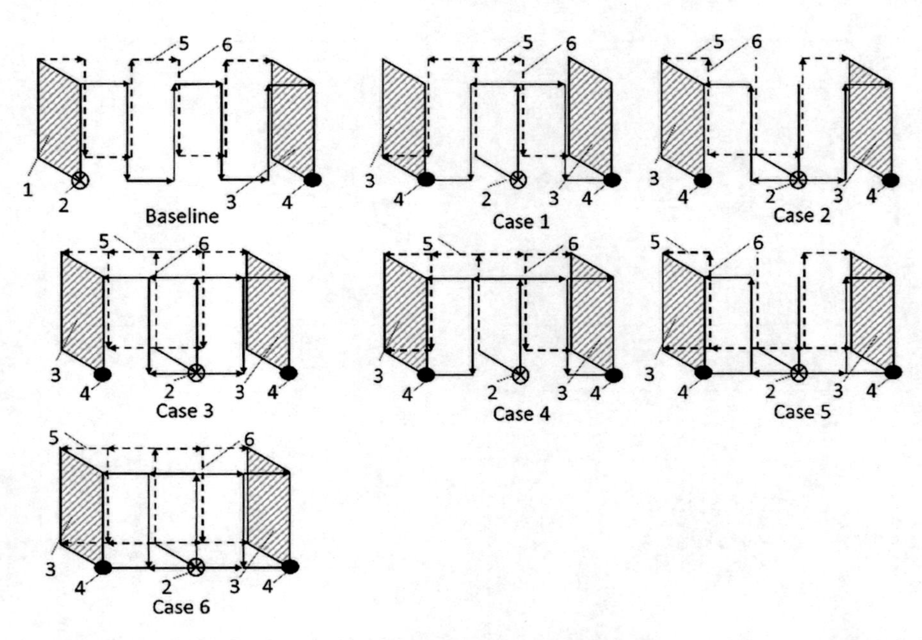

Fig. 18.2 Flow channel diagram of each arrangement with on the short side. 1-inlet sink, 2-inlet, 3-outlet sink, 4-outlet, 5-transverse flow channel, 6-longitudinal flow channel, with the arrows indicating the flow directions and shadowed regions of the plenum

$$f_D = \left(d_{T_{\max}}^{r_1} \cdot d_T^{r_2} \cdot d_P^{r_3}\right)^{\frac{1}{\sum r_i}} \tag{18.1}$$

$$d_{T\max} = \frac{31 - T}{31 - 25}, 25 < T < 31 \tag{18.2}$$

$$d_T = \frac{5 - \Delta T}{5 - 0}, 0 < \Delta T < 5 \tag{18.3}$$

$$d_P = \frac{2000 - \Delta P}{2000 - 0}, 0 < \Delta P < 2000 \tag{18.4}$$

18.3 Results and Discussion

18.3.1 Effect of Different Charge/Discharge Multipliers

In this section, the variation of battery module temperature at different charge/discharge multipliers and different flow rates is experimentally investigated.

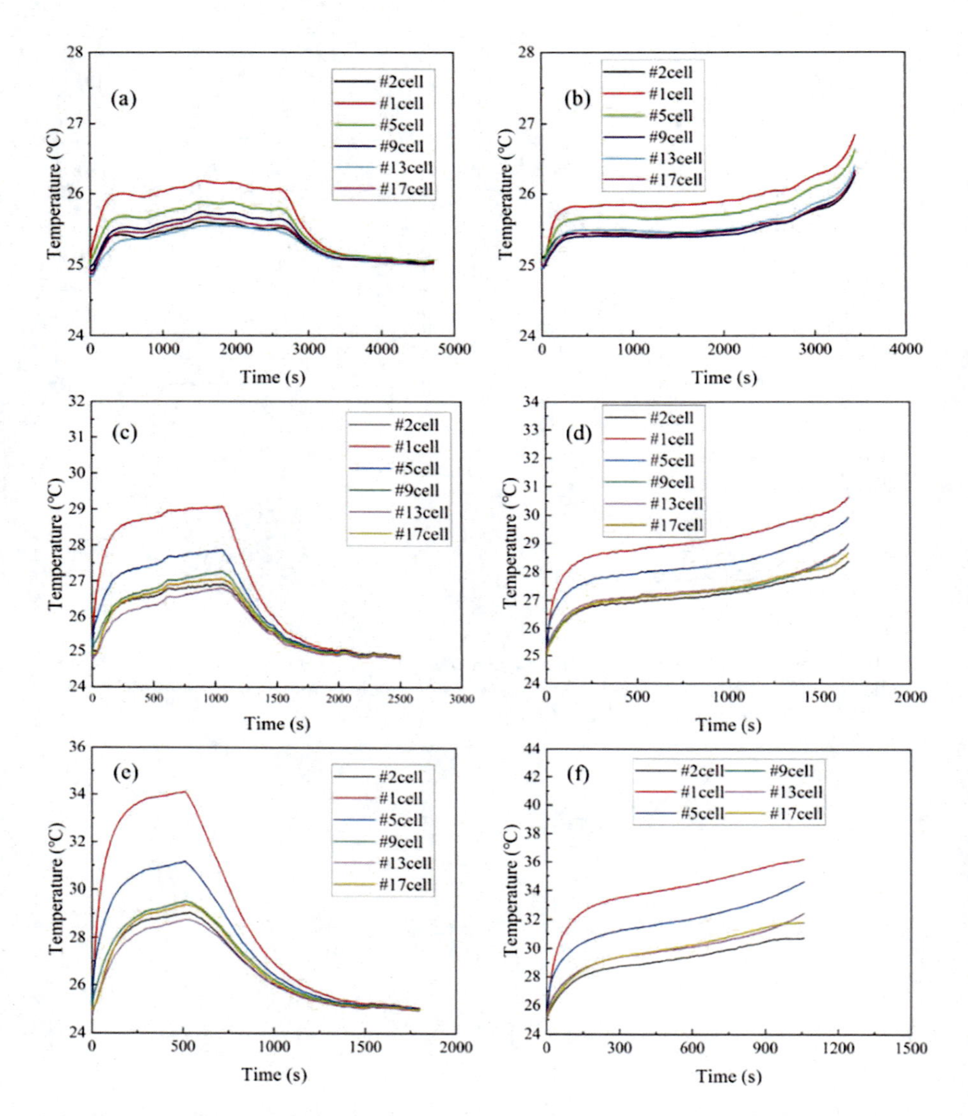

Fig. 18.3 The change of battery module temperature with different discharge and charge rates. (**a**) 1C charge, (**b**) 1C discharge, (**c**) 2C charge, (**d**) 2C discharge, (**e**) 3C charge, (**f**) 3C discharge

The variation of temperature with discharge time for 18,650 battery modules at different charge and discharge multipliers is shown in Fig. 18.3. The maximum temperature of the battery module at 3C discharge multiplier in this experiment is 36.21 °C, which is lower than the threshold of battery operating temperature, and the lowest temperature of the #2 battery is also the lowest temperature of the module, which is 30.73 °C, and the temperature difference of the battery module is 5.48 °C.

The maximum temperature of the battery module at 3C charging multiplier is 34.1 °C, which is lower than the maximum temperature of the discharge at the same multiplier, and the lowest temperature of the #2 battery is also the lowest temperature of the module at 28.74 °C, and the temperature difference of the battery module is 5.36 °C. It can be seen that the new liquid-cooled shell structure has good heat dissipation and temperature equalization performance in the battery charging and discharging process.

18.3.2 Effect of Different Liquid Cooling Flow Rates

The variation of cell module temperature, temperature difference, and inlet/outlet pressure drop with coolant flow rate is shown in Fig. 18.4. The cell was discharged at a rate of 3C during the experiment. The flow rates were tested in the order of 20 L/h (0.2 m/s), 30 L/h (0.3 m/s), 50 L/h (0.5 m/s), 70 L/h (0.7 m/s), and 100 L/h (1.0 m/s). It can be seen from the Fig. 18.4 that the temperature change trend of the battery module is the same for different flow rates at 3C discharge multiplier. As the flow rate increases, the maximum temperature of the battery module gradually decreases. Nonetheless, the temperature difference does not change significantly, whereas the pressure drop keeps increasing and so does the power consumption of the pump. Therefore, the use of a high flow rate is not cost-effective for the new heat dissipation structure. Instead, a small flow rate around 0.3 m/s ~ 0.5 m/s is sufficient to meet the temperature requirements of the battery module structure.

18.4 Conclusion

In this paper, we study a new battery module, which consists of 4×5 cells in a liquid-cooled shell structure with different flow configurations. The conclusions are summarized as follows:

1. The thermal performance of the baseline case and six types of flow structures were studied. The expectation function was employed to identify the optimal flow configuration of Case 1, which reduces the pressure drop by 66.5% while the maximum temperature and temperature difference are kept around the lowest levels.
2. The highest temperature of the battery module increases with the increase of the discharge rate or charge rate, and the temperature of discharge is higher than the temperature of charge at the same rate. The greater the discharge rate is, the greater the degree of influence of the heat of the busbar is.

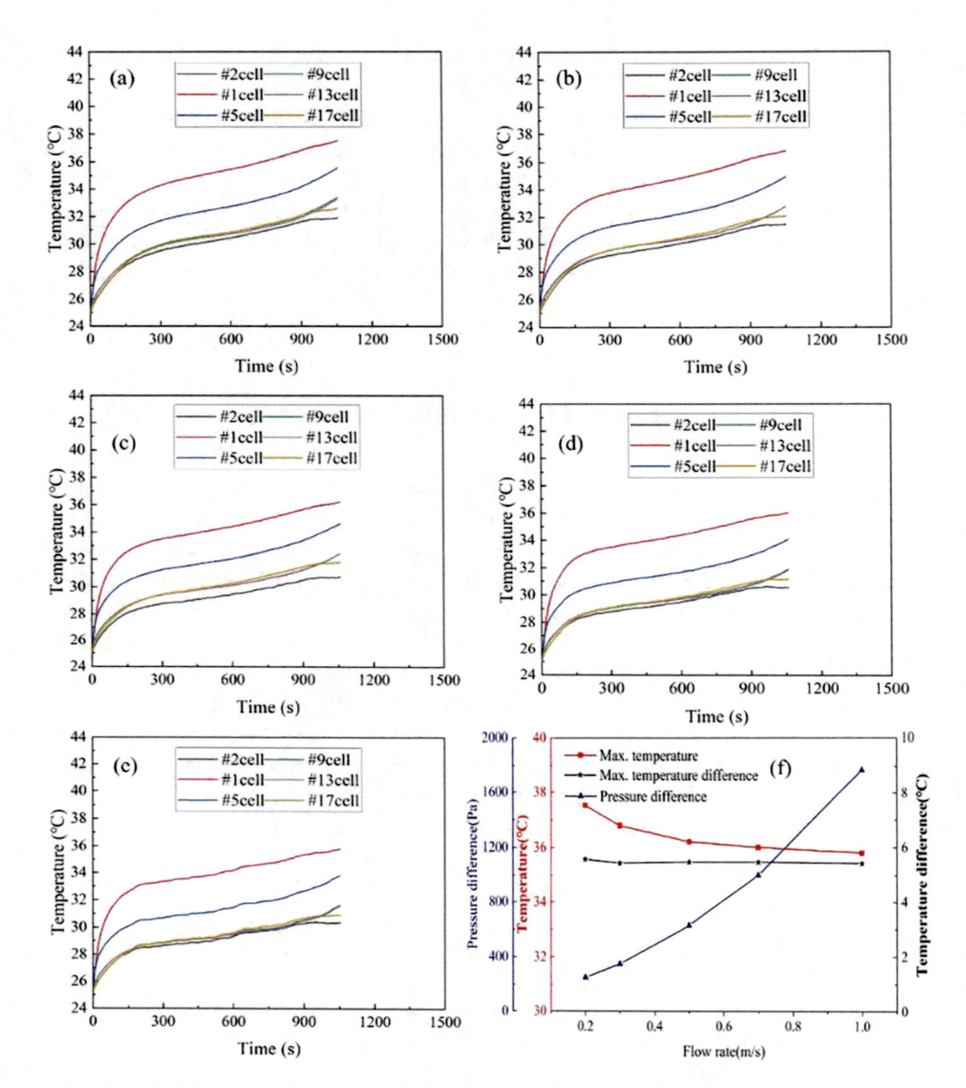

Fig. 18.4 The change of battery module temperature with discharge time at different flow rates. (**a**) 0.2 m/s, (**b**) 0.3 m/s, (**c**) 0.5 m/s, (**d**) 0.7 m/s, (**e**) 1.0 m/s and (**f**) Tmax, temperature difference and pressure drop vs the flowrate

References

1. Patel, J.R., Rathod, M.K.: Recent developments in the passive and hybrid thermal management techniques of lithium-ion batteries [J]. J. Power Sources. **480**, 228820 (2020)
2. Aris, A.M., Shabani, B.: An experimental study of a lithium ion cell operation at low temperature conditions [J]. Energy Procedia. **110**, 128–135 (2017)

3. Gao, Y., Jiang, J.C., Zhang, C.P., et al.: Lithium-ion battery aging mechanisms and life model under different charging stresses [J]. J. Power Sources. **356**, 103–114 (2017)
4. Onda, K., Ohshima, T., Nakayama, M., et al.: Thermal behavior of small lithium-ion battery during rapid charge and discharge cycles [J]. J. Power Sources. **158**(1), 535–542 (2005)
5. Rao, Z.H., Wang, S.F.: A review of power battery thermal energy management [J]. Renew. Sust. Energ. Rev. **15**, 4554–4571 (2011)
6. Lindgren, J., Lund, P.D.: Effect of extreme temperatures on battery charging and performance of electric vehicles [J]. J. Power Sources. **328**, 37–45 (2016)
7. Chen, K., Wu, W.X., Yuan, F., et al.: Cooling efficiency improvement of air-cooled battery thermal management system through designing the flow pattern [J]. Energy. **167**, 781–790 (2019)
8. Choudhari, V.G., Dhoble, A.S., Satyam, P.: Numerical analysis of different fin structures in phase change material module for battery thermal management system and its optimization [J]. Int. J. Heat Mass Transf. **163**, 120434 (2020)
9. Zhao, C.R., Sousa Antonio, C.M., Jiang, F.M.: Minimization of thermal non-uniformity in lithium-ion battery pack cooled by channeled liquid flow [J]. Int. J. Heat Mass Transf. **129**, 660–670 (2019)

Chapter 19
A Study on Heat Generation of Lithium-Ion Battery Used in Electric Vehicles by Simulation and Experiment

K. Selvararajoo, V. Vicki Wanatasanappan, and N. Y. Luon

19.1 Introduction

Electric vehicle (EV) is developing as it is one of the transportation that does not involve any fossil fuel combustion process. Most importantly, it is beneficial to the environment by reducing air and noise contamination. Electric vehicles use high energy batteries to provide sufficient charge for them to run. Electric vehicles reduce the impact on the environment and also enhance energy performance. So, EVs and HEV were developed to encourage sustainable development in the transportation field [1]. Lithium-ion battery (LIB) is one of the high energy batteries that meet the criteria because of their good reliability, rich energy density and low self-discharge, zero memory power, lower weight and long lifespan. Nevertheless, LIB has disadvantages such as thermal instability or thermal runway and degradation of capacity [2]. The performance of the battery and its lifetime is depending on its working condition [3]. When the battery charges and discharges, the exothermic process takes place and releases heat. This chemical combustion reaction may cause an increase in the battery temperature, specifically at the end of the cycle. The heat release in the battery will cause harm to the battery itself and for the passengers in form of safety and reliability when utilizing in EV as reported by Mohammad et al. [4]. The battery function can be extremely poor at very low and high temperatures [3]. Due to its sensitivity to temperature rise, it might lead to a fire explosion due to electrolyte fire when the battery gets overheated [5].

K. Selvararajoo
College of Graduate Studies, Selangor, Malaysia

V. V. Wanatasanappan (✉) · N. Y. Luon
Institute of Power Engineering, Universiti Tenaga Nasional, Selangor, Malaysia
e-mail: vignesh@uniten.edu.my

X. Wang (ed.), *Future Energy*, Green Energy and Technology,
https://doi.org/10.1007/978-3-031-33906-6_19

The ideal temperature range for operating the LIB is between 25 and 40 °C [6, 7]. This is because the battery shelf life will decrease by 60 days as the temperature increases at each degree Celsius mentioned by R. Zhao et al. [8]. The maximum temperature is restricted up to 60 °C for the thermal management system of the battery, and the charging amount should be constant. The temperature variance is controlled at less than 5 °C to support the battery balance [9]. Thus, a battery heat management system is crucial for maintaining a constant temperature and meeting the required range. Furthermore, the battery must operate at a required temperature range in order to save energy and the cost of electric vehicles [3]. Chunrong Zhao reported that to enhance the operation method and to maintain the uniform maximum temperature, the battery design, dimensions and arrangements need to be considered. The author mentioned that the module should not contain a large number of batteries [2].

When focusing on the cylindrical type LIB, Mikael et al. [10] reported that the maximum heat generates at the core, which is at the vertical centre area of the battery, spread to the surface of the battery cell. This is due to the formation of ions when charging and discharging. The Li-ion will extrapolate from the positive electrode and then interpolate into negative electrodes during when charging process, and the formation will be opposite for discharging process. Ke Li et al. [9] reported an example of ion formation during the charging and discharging process as shown in Eqs. 1 and 2. The heat generation of the cylindrical LIB cell is observed by simulation and experimental work in this project. Panasonic NCR 18650B battery cell with the dimension of $D = 18.5$ mm and $h = 65.0$ mm is selected and the battery system circuit is designed using MATLAB Simulink model. The parameter of battery specification and the charging/discharging current rate is set into the software to simulate a temperature rise graph against the state of charge (SOC) and depth of discharge (DOD). For experimental work, the battery setup was prepared with a thermocouple to conduct the charging/discharging process. The temperature is captured by thermocouples connected to the PicoLog TC 08 data logger.

19.2 Methodology

19.2.1 LIB Simulation

In the present study, a predefined battery system circuit was constructed using the MATLAB Simulink model to get stable output. Three models, which are the charging/discharging battery system, subsystem and data collection system, were designed as shown in Figs. 19.1, 19.2 and 19.3. The input parameter values to run the simulation were filled as shown in Table 19.1. The parameters were filled according to NCR 18650B battery specification from the case data released by Panasonic, which is tabulated in Table 19.2. To charge and discharge the battery at the different current rates of 0.5C, 1C and 3C, the current value was set for each current rate

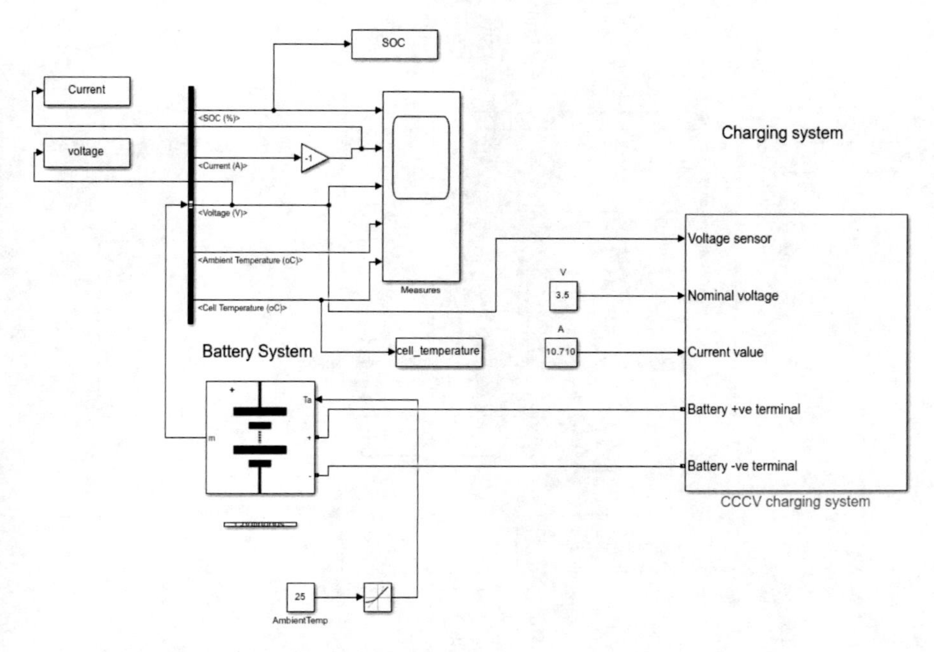

Fig. 19.1 Charging system of Li-ion battery

based on Table 3. The ambient temperature was set to 25 °C. The simulation was run for charging and discharging at 0.5C, 1C and 3C rates, respectively, and the individual temperature rise graph is plotted.

19.2.2 Experimental Procedure for LIB Charging and Discharging

Figure 19.4a, b illustrates the experimental setup to investigate the temperature rise of LIB during the charging and discharging process. The Panasonic NCR 18650 Li-ion battery with 3300 mAh capacity is used for the experiment. Two K-type thermocouples were attached to the LIB surface to monitor the temperature rise. The LIB with thermocouples attached were insulated with super wool to prevent heat loss to the surrounding during the experimental procedure. To monitor the ambient temperature, one thermocouple was used to record the surrounding temperature data. The temperature data are logged through the Pico TC08 data logger, which has temperature sensitivity up to ±0.1 °C. For the charging process, IMAX B8 battery charger was used; meanwhile for the constant current discharging process, a 150 W electronic load tester was used. Three different constant current discharge rates of 0.5C, 1C and 3C were investigated for each experiment.

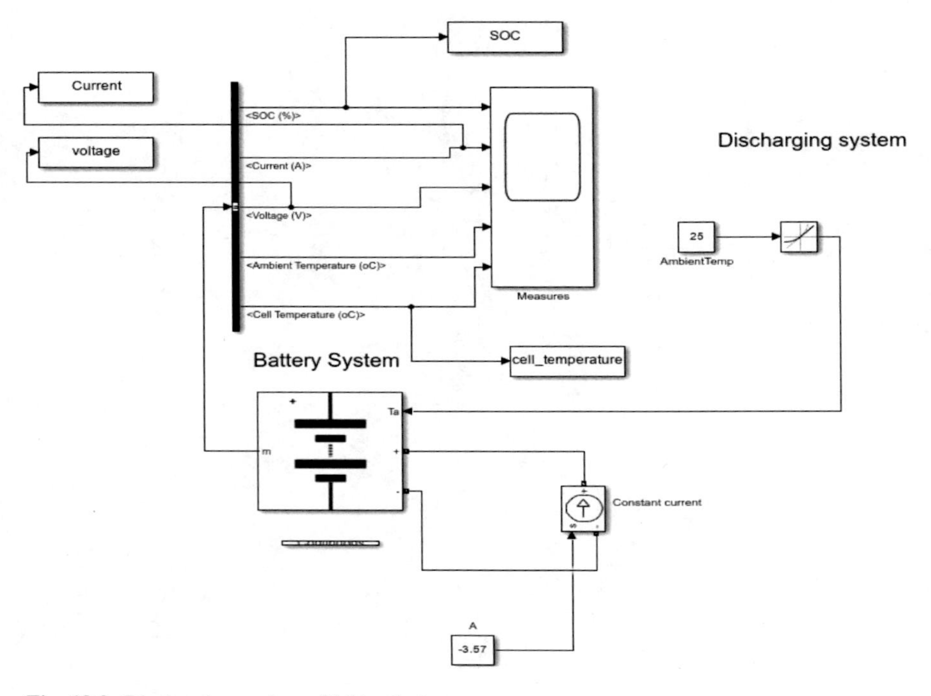

Fig. 19.2 Discharging system of Li-ion battery

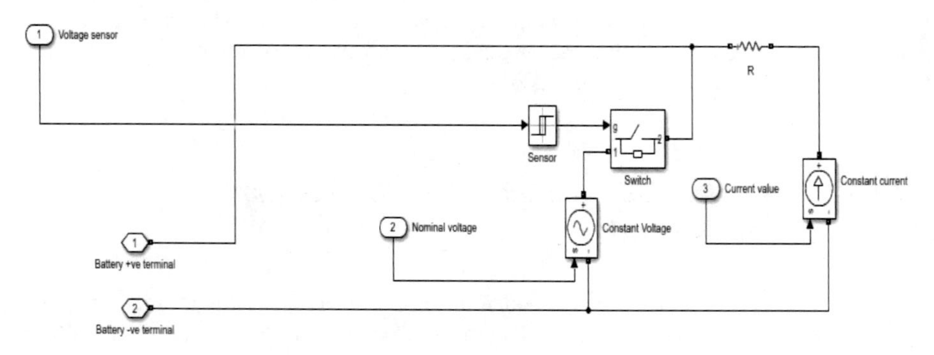

Fig. 19.3 Li-ion battery subsystem

19.3 Results and Discussion

19.3.1 Effect of Charging Process on LIB Temperature

Figure 19.5 shows the temperature rise of LIB against the state of charge for the charging process at 1C by simulation; meanwhile, the temperature rise of LIB during the charging process by the experimental procedure is displayed in Fig. 19.6. The

Table 19.1 Parameter of predefined battery

Parameters	Values
Initial cell temperature (°C)	25
Nominal ambient temperature (°C)	25
Second nominal ambient temperature (°C)	30
Maximum capacity (Ah)	3.57
Initial discharge voltage (V)	3.6
Capacity (Ah) at nominal voltage	3.35
Voltage at 90% maximum capacity (V)	2.5
Exponential zone [voltage (V), capacity (Ah)]	3.7, 0.583
Thermal resistance, cell-to-ambient (oC/W)	0.6
Thermal time constant, cell-to-ambient (s)	2000
Heat loss difference [charging vs discharging] (W)	0

Table 19.2 Specification Panasonic NCR 18650B battery

Specification	Range
Capacity	Min. 3250 mAhTyp. 3350 mAh
Nominal voltage	3.6 V
Charging	CC-CV, Std. 1625 mA, 4.20 V, 4.0 h
Weight (max).	48.5 g
Temperature	Charge*: 0 to +45 °CDischarge: −20 to +60 °C
Energy density	Volumetric: 676 Wh/lGravimetric: 243 Wh/kg

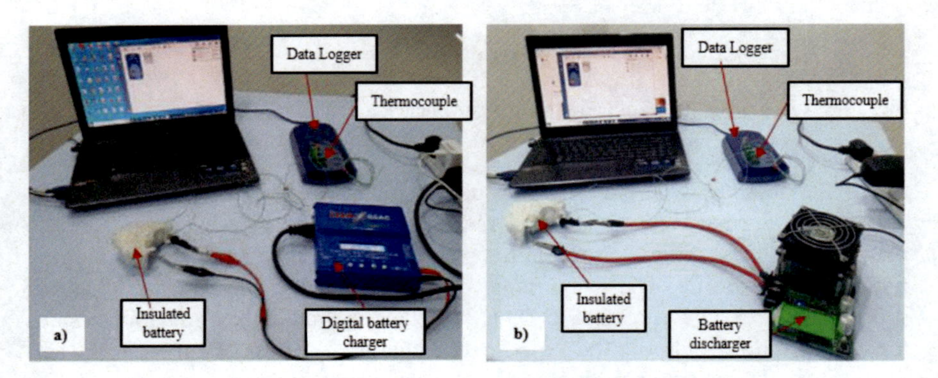

Fig. 19.4 Experimental setup (**a**) Charging and (**b**) discharging process

presented results show that the temperature of LIB rises almost linearly during the charging simulation. The highest temperature recorded during 1C charging simulation is 304.75 K. Once the temperature reaches the peak and SOC is 100%, the charging current becomes zero and the temperature falls instantaneously to 300.66 K. However, the temperature profile of LIB cell during the experimental charging process shows slightly different behaviour. The LIB temperature increased to a maximum value of 304.56 K at SOC of 40%. As the SOC increases further,

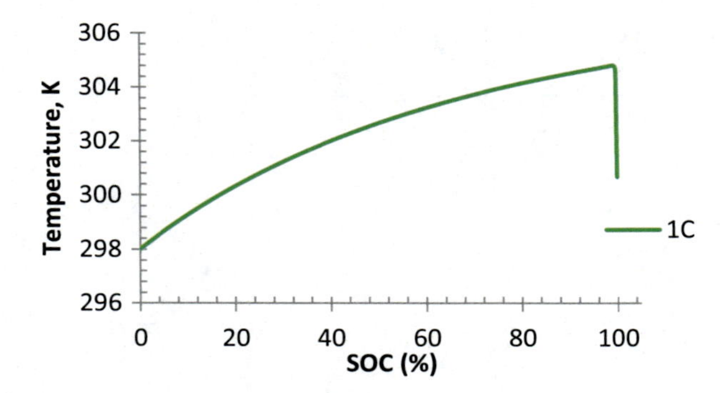

Fig. 19.5 LIB temperature profile during simulation charging

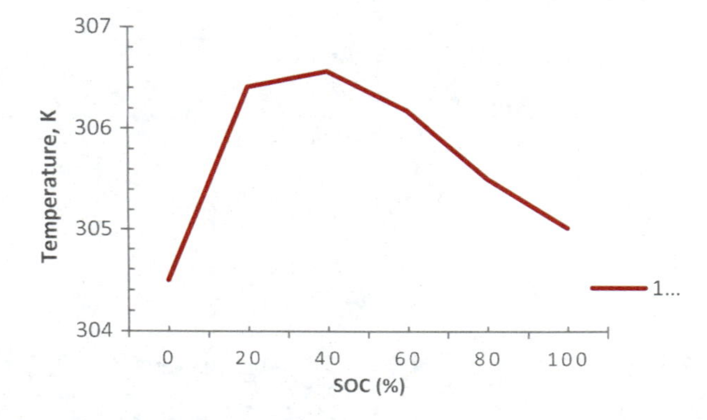

Fig. 19.6 LIB temperature profile during experimental charging

the LIB temperatures started to show a decreasing trend. This is mainly due to the reduction in current value as the SOC reaches almost 50%. However, compared to the simulation charging, the drop in temperature during the experimental charging process is not drastic. The temperature of LIB during the experimental charging process at 100% SOC is about 305.03 K. When compared to the simulation results, the temperature rise during the experimental charging process is only 0.38%. This is significantly lower when compared to the temperature rise during simulation charging, which is about 2.17%.

When comparing both graphs, the simulation graph curves show a smooth temperature rise and the temperature is directly proportional to the SOC. A similar pattern of temperature rise is found in a study by Ke Li et al. [9] and there stated that the heat formation is reduced due to the changes from the charge of uniform current to the charge of uniform voltage. As increasing the charging duration, the negative electrode became impregnated and leads the battery resistance to rise. Therefore, some researchers proposed to lower the charging current once constant voltage took over, which is unable to keep charging for a longer time and produces more heat in the battery [9, 11, 12].

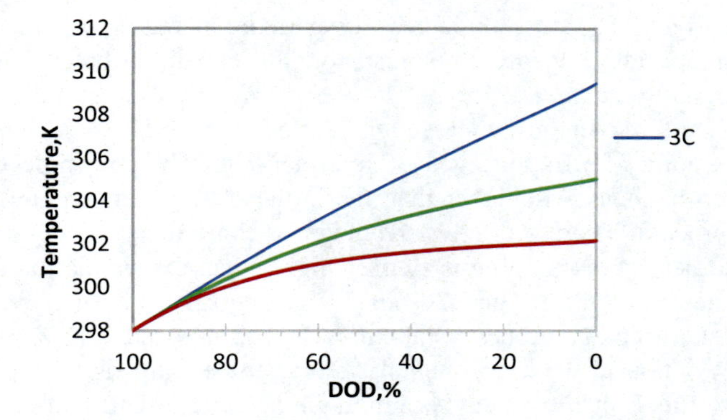

Fig. 19.7 LIB temperature profile during simulation discharging

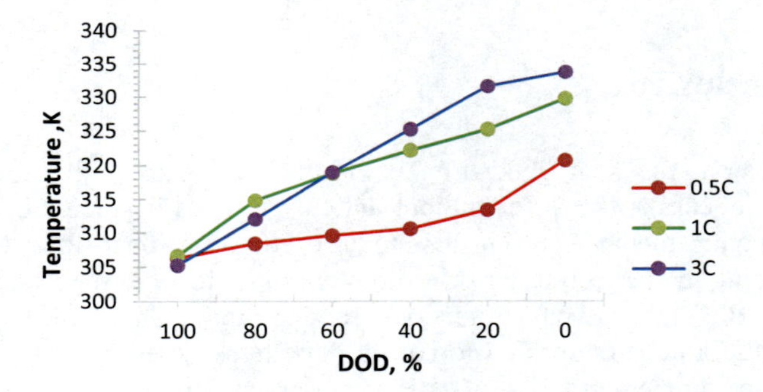

Fig. 19.8 LIB temperature profile during experimental discharging

19.3.2 Effect of Discharging Process on the LIB Temperature

Figures 19.7 and 19.8 show the graph of LIB temperature distribution against the depth of discharge during discharging process at 0.5C, 1C and 3C for both simulation and experiment, respectively. The simulation graph for discharging process revealed that the temperature of LIB increases as the DOD increases from 100 to 0% for all three discharge rates simulated. The increase in battery temperature is due to the constant current discharge of the LIB. The rise in battery temperature is greater for a higher discharge rate. For instance, the highest battery temperature of 309.44 K is recorded for a 3C discharge rate, and the rise in battery temperature is about 3.84%. For a discharge rate of 0.5C and 1C, the increment in battery temperature is about 1.39% and 2.31%, respectively. Moreover, the rise in battery temperature for a higher discharge rate of 3C is almost linear compared to the 0.5C and 1C discharge rates, which exhibit a smooth curve of rise in LIB temperature.

The experimental results obtained for discharging LIB at 0.5 C, 1C and 3C discharge rates exhibit an almost identical trend as the simulation results. As displayed in Fig. 19.8, the rise in battery temperature increases significantly as the battery discharge rate increases. For instance, the maximum battery temperature obtained for a discharge rate of 3C is 333.79 K; meanwhile, for a discharge rate of 0.5C, the maximum battery temperature is only 320.78 K. However, the initial temperature of LIB during the experimental discharging process is about 305 K, which is about 7 K higher than the initial battery temperature during the simulation process. Therefore, the percentage rise in battery temperature for the experimental battery discharging is slightly higher than the simulation results. For discharge rates of 0.5C, 1C and 3C, the percentage rise in battery temperature for a complete discharge process is about 5.17%, 8.11% and 9.42%, respectively. The slight variation in the experimental battery temperature rise compared to the simulation results is due to the higher initial temperature of LIB that contributed to the higher kinetic energy of molecules at the initial stage.

19.4 Conclusion

In this research, the heat generation of 18,650 LIB during the charging and discharging process at the different constant current rates is studied by numerical and experimental methods. From this study, it could be found that for charging and discharging the temperature rise is directly proportional to the SOC and DOD, respectively. But the temperature increment for discharging during the experiment is higher than the simulation result. Moreover, for both the experiment and simulation, the maximum temperature rise of LIB is observed for 3C discharge rate. The battery generates more heat when discharged during the experiment, which is a real-time condition. For charging, the simulation result temperature is greater than an experiment. Besides, the minor variation in the experimental results is due to the ambient temperature and initial temperature of the battery during both the experiment and simulation.

Acknowledgements The authors acknowledge Universiti Tenaga Nasional for the financial support through YCU 2022 research grant (202210032YCU). Special thanks to those who contributed to this project directly or indirectly.

References

1. Al-Zareer, M., Dincer, I., Rosen, M.A.: A review of novel thermal management systems for batteries. Int. J. Energy Res. **42**, 3182–3205 (2018)
2. Zhao, C., Cao, W., Dong, T., Jiang, F.: Thermal behavior study of discharging/charging cylindrical lithium-ion battery module cooled by channeled liquid flow. Int. J. Heat Mass Transf. **120**, 751–762 (2018)

3. Putra, N., Ariantara, B., Pamungkus, R.A.: Experimental investigation on performance of lithium-ion battery thermal management system using flat plate loop heat pipe for electric vehicles application. Appl. Therm. Eng. **99**, 787–789 (2016)
4. Shahjalal, M., Tamanna Shams, M., Islam, E., Alam, W., Modak, M., Hossai, S.B., Venkatasailanathan Ramadesigan, M., Ahmed, R., Ahmed, H., Iqbal, A.: A review of thermal management for Li-ion batteries: prospects, challenges, and issues. J. Energy Storage. **39**, 102518 (2021)
5. Lu, L., Han, X., Li, J., Hua Ouyang, M.: A review on the key issues for lithium-ion battery management in electric vehicles. J. Power Sources. **88**, 226–272 (2013)
6. Qian, Z., Li, Y., Rao, Z.: Thermal performance of lithium ion battery thermal management system by using mini-channel cooling. Energy Conser. Manage. **126**, 622–631 (2016)
7. Pesaran, A.A.: Battery thermal models for hybrid vehicle simulations. J. Power Sources. **110**, 377–382 (2002)
8. Zhao, R., Gu, J., Liu, J.: An experimental study of heat pipe thermal management system with wet cooling method for lithium ion batteries. J. Power Sources. **273**, 1089–1097 (2015)
9. Li, K., Yan, J., Chen, H., Wang, Q.: Water cooling based strategy for lithium ion battery pack dynamic cycling for thermal management system. Appl. Therm. Eng. **132**, 575–585 (2018)
10. Katrasnik, T., Mele, I., Zelic, K.: Multi-scale modelling of Lithium-ion batteries: From transport phenomena to the outbreak of thermal runaway. Energy Conver. Manage. **236**, 114036 (2021)
11. Menale, C., D'Annibale, F., Mazzarotta, B., Bubbico, R.: Thermal management of Lithium-ion batteries: an experimental investigation. Energy. **182**, 57–71 (2019)
12. Jialong, L., Zonghou, H., Jinhua, S., Qingsong, W.: Heat generation and thermal runaway of lithium-ion battery induced by slight overcharging cycling. J. Power Sources. **526**, 231136 (2022)

Part V
Fuel Consumption, Transportation Carbon Emissions and Climate Change Management

Chapter 20
Prediction of Fuel Consumption of Heavy Commercial Vehicles Based on Random Forest

Shi Guodong, Fang Jian, Hu Mingmao, and Xiang Haijing

20.1 Introduction

At present, the world is actively developing new energy vehicles, but in a short period of time, new energy vehicles cannot replace fuel vehicles. Therefore, it is urgent to analyze and study the fuel consumption of vehicles, so as to reduce fuel consumption and vehicle exhaust emissions; it can also reduce vehicle costs and improve fuel economy. Studies have shown that driving a vehicle under different operating conditions can lead to significant differences in fuel consumption.

At present, there have been many researches in the field of vehicle fuel consumption prediction. Huang He et al. [1] built a fuel consumption prediction model based on support vector regression (SVR) and multilayer perceptron (MPL) and collected vehicle operation data and analyzed the relationship between various driving parameters of the vehicle and fuel consumption. Yang Yalian et al. [2] combined generalized regression neural network with dynamic programming algorithm to build a fuel consumption prediction model and compared it with the DP simulation results to verify the effectiveness of the model. Su Xiaohui et al. [3] used the improved K-means clustering algorithm to study the driving conditions and fuel consumption of automobiles and solved the situation that the traditional clustering algorithm is easy to fall into the local optimal solution and takes a long time. Wickramanayake et al. [4] established three long-distance bus fuel consumption

S. Guodong · H. Mingmao (✉)
School of Mechanical Engineering, Hubei Institute of Automotive Industry, Zhangwan District, Shiyan, Hubei, China

F. Jian
Dongfeng Commercial Vehicle Co., Ltd, Hubei, China

X. Haijing
Dongfeng Equipment Manufacturing Co., Ltd, Hubei, China

X. Wang (ed.), *Future Energy*, Green Energy and Technology ,
https://doi.org/10.1007/978-3-031-33906-6_20

prediction models based on neural networks, gradient boosting trees, and random forests, respectively, according to the two different road slopes of vehicles going up and down. Based on the time series of vehicle speed, acceleration, and road gradient as input, Topić Jakov et al. [5] built a fuel consumption prediction model with linear regression and neural network. The results show that the proposed method based on neural network provides good prediction accuracy and reasonable accuracy. Capraz et al. [6] used multiple linear regression, neural network, and support vector machine regression (SVR), which are the three algorithms to predict vehicle fuel consumption and total fuel consumption. Yang Lili et al. [7] built a fuel consumption model of grain combine harvesters based on random forest and analyzed the influence of seven indicators, such as the average engine torque and average engine speed of grain harvesters, on the oil consumption of the harvester under different geographical backgrounds in my country. Operating conditions and fuel consumption optimization provide theoretical basis.

In this paper, using python language as a tool, based on the real-time running state data of vehicles collected from the T-BOX interface of a certain type of heavy-duty commercial vehicle, two regression models of vehicle fuel consumption are built using random forest. The relationship between operating conditions provides a theoretical basis to a certain extent.

20.2 Data Source and Preprocessing

20.2.1 Data Sources

The data in this paper comes from the Internet of vehicles system of a commercial vehicle company's heavy-duty commercial vehicles. The data set of the vehicle's running state in a short period of time is obtained through the onboard T-BOX in real time. The data sample is large and comprehensive, with universality and diversity. The parameters of the original data set include fuel consumption, longitude, latitude, GPS speed, engine torque, engine speed, accelerator opening, oil pressure, engine water temperature, air temperature, brake switch status, and air conditioner switch status. The fuel consumption is the target value, and the remaining parameters are for the feature values.

20.2.2 Data Preprocessing

It can be seen from Table 20.1 that the data sampling frequency is 1s, which is a high sampling frequency. If this high sampling frequency data is directly input into the model without processing, it is easy to cause the model to overfit. Considering the actual working conditions, study within 1s. The actual meaning of vehicle fuel

Table 20.1 Optimal parameters of the grid-based search

	n	h	N	m
Initial parameters	100	15	1	2
Optimal parameters	81	9	2	2

consumption in a very short time is not great, so the collected data are grouped in order according to the time step of 5s, and the average value of each time step is taken as the research object, which can prevent the model from overfitting and can better reflect the driving conditions of the vehicle.

The statistical feature formula of downsampling is as follows:

$$f(x) = \frac{1}{T}\sum_{i=1}^{T} x_i \tag{20.1}$$

In the formula, T is the downsampling period, xi is the corresponding characteristic parameter, and $f(x)$ represents the downsampling result obtained when xi takes different values.

1. The outliers are replaced by interpolation, and the interpolation adopts the following methods:

$$l_t = (l_{t+1} + l_{t-1})/2 \tag{20.2}$$

where l_t represents the abnormal eigenvalue at time, $l(t + 1)$ is the normal eigenvalue at $t + 1$ time, and $l(t - 1)$ is the normal eigenvalue at $t - 1$ time.
2. Standardize the data, and linearly transform the data to make the original data close to a normal distribution with a mean of 0 and a variance of 1. The data normalization formula is as follows:

$$X' = \frac{X - u}{\sigma} \tag{20.3}$$

where $X^{'}$ represents the standardized value, X represents the initial value, u represents the feature mean, and σ represents the feature standard deviation.

20.3 Analysis and Extraction of Basic Feature Parameters

20.3.1 Feature Parameter Extraction

In order to reduce the amount of model calculation, 12 features need to be screened, the SelectKbest function in the SKlearn library is called, and the threshold n is set to 10; that is, 10 optimal parameters are selected from the 12 feature parameter values and then the get_support( ) method to view the selected 10 feature parameter names.

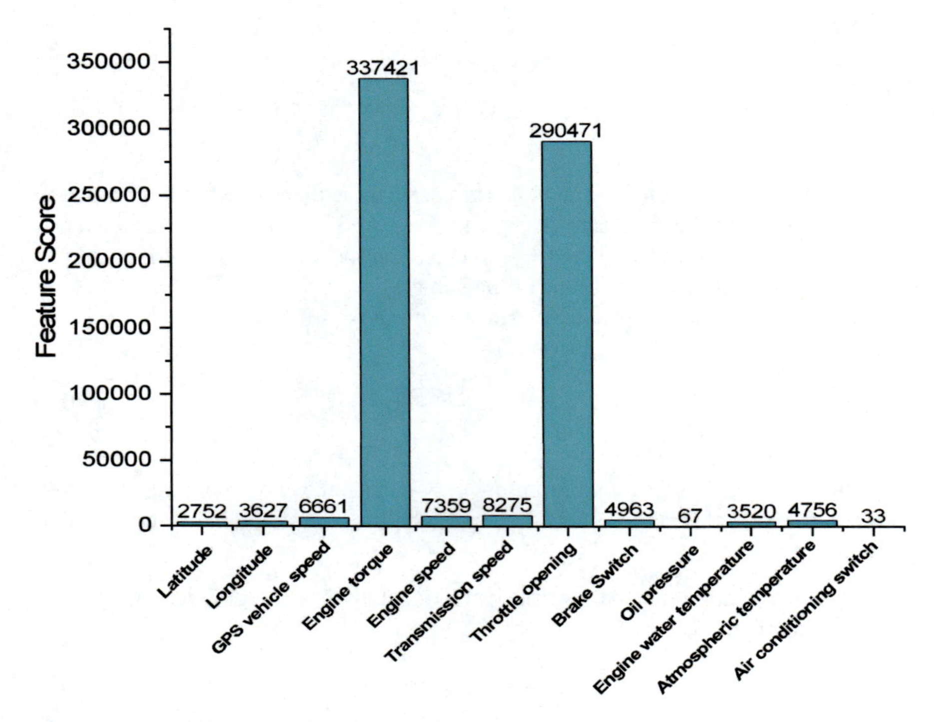

Fig. 20.1 Feature parameter score

The score of each feature parameter screened out by the SelectKbest function is shown in Fig. 20.1.

20.4 Construction and Result Analysis of Fuel Consumption Prediction Model

20.4.1 Construction of Random Forest Fuel Consumption Prediction Model

1. Model building

This paper builds a random forest fuel consumption prediction model based on Python 3.6 and SKlearn 0.19.1 machine learning library. The ten feature parameters selected in Sect. 20.2.1 are standardized as the input features of fuel consumption prediction, 75% of the data is divided into training set, 5814 training samples are obtained, and the remaining 25% is divided into test set, and 1939 test samples are obtained. In order to prevent the model from overfitting, tenfold cross-validation

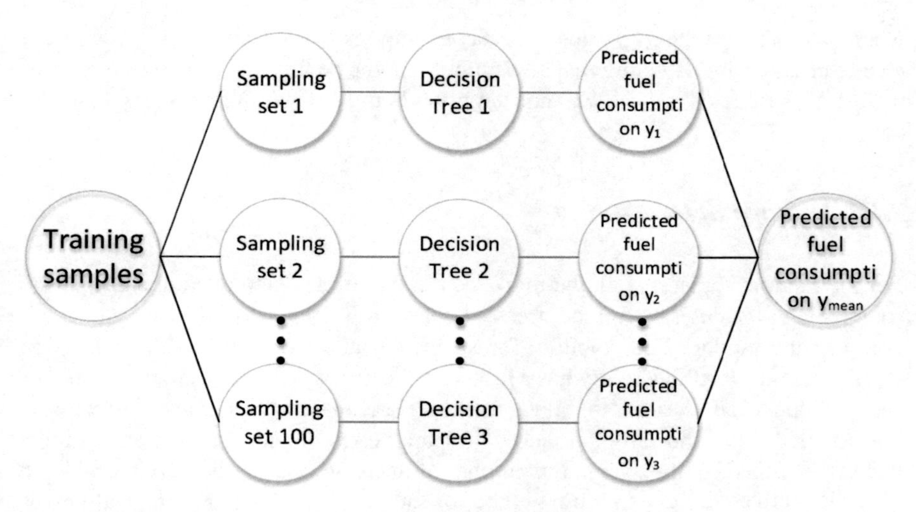

Fig. 20.2 Structure of random forest fuel consumption prediction model

was performed on the training set, and grid search was used to find the optimal parameters of the random forest model.

Random forest is a typical comprehensive learning algorithm, which is the representative of bagging method. Its weak learner is the Classification and Regression Trees (CART) decision tree model. The initial number of decision trees n of the model is given as 100, the maximum depth h of the decision tree is given as 15, the minimum number of samples N contained in the leaf nodes is given as 1, the minimum number of samples m that nodes can be divided into is given as 2, and the rest parameters are adopted. The default value of the model, The structure of the random forest fuel consumption prediction model is shown in Fig. 20.2.

As seen in Fig. 20.2, 100 sampling sets were constructed by random sampling of training samples, and then 100 decision trees were constructed as weak learners for the fuel consumption prediction model based on these 100 sampling sets [7], where each node of the decision tree contains n training samples, and a feature subset $X^{'}$ consists of k training samples randomly, and node splitting is performed through the feature subset $X^{'}$. Each node can be split into two new nodes $R_1(q, s)$ and $R_2(q, s)$, where q represents the relevant indexes that have an impact on fuel consumption and $X^{'}$ represents the node splitting threshold, the objective function of the following method:

$$\underbrace{\min}_{q,s}\left[\underbrace{\min}_{c1}\sum_{x_l\in R_1(q,s)}(y_l-c_1)^2+\underbrace{\min}_{c_2}\sum_{x_l\in R_2(q,s)}(y_l-c_2)^2\right] \tag{20.4}$$

where y_l represents the true value of fuel consumption in the l entry, x_l represents the characteristic value affecting fuel consumption in the l entry, and c_1 and c_2 represent the average value of the true fuel consumption in the R_1 and R_2 nodes, respectively.

20.4.2 Parameter Tuning

For the random forest model, the more the number of decision trees, the deeper the root depth of the decision number trees, and the more branches of the trees, the more complex the model. Since random forest itself is an algorithm based on decision trees, random forest inherently has a high complexity, which is not conducive to our model calculation. In order to reduce the model complexity and to avoid overfitting due to small bias and large variance, the parameters of the random forest model need to be tuned. In this paper, the number of decision trees n, the maximum depth of decision trees h, the minimum number of samples N contained in the leaf nodes, and the minimum number of samples m that can be divided by nodes are searched in a grid, and the optimal parameters are obtained in Table 20.1.

As can be seen from Table 20.1, when the number of decision trees is taken as 81, the maximum depth of decision tree h is taken as 9, the minimum number of samples N contained in the leaf nodes is taken as 2, and the minimum number of samples m that can be divided by the nodes is taken as 2, the model is the optimal model, and the model predicts the highest accuracy.

The regression model was rebuilt using the optimal parameters, and the fuel consumption prediction error was reduced by 0.001, and the optimal model score is shown in Fig. 20.3.

As can be seen from Fig. 20.3, the model score of the random forest fuel consumption prediction model gradually stabilizes when the number of cross-validation samples reaches about 5000, and the final score is fixed at 0.8, which shows that the model trains faster and the model prediction accuracy is higher.

20.4.3 Verification and Analysis of Fuel Consumption Prediction Model Results

In order to verify whether the above two fuel consumption prediction models can meet the requirements, the results need to be analyzed and calculated. In this paper, three different indicators are used to evaluate and analyze the models, namely, mean square error (MSE), mean absolute error (MAE), and R_2 coefficient, and the three indicators are calculated in the following way:

$$\text{MSE} = \frac{1}{m}\sum_{i=1}^{m}\left(y_i - \hat{y}_i\right)^2 \tag{20.5}$$

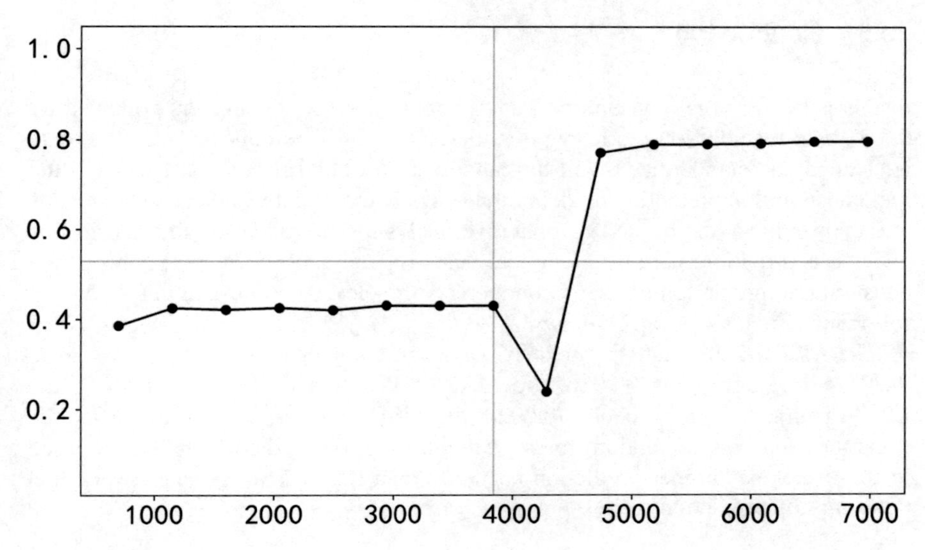

Fig. 20.3 Optimal model score

Table 20.2 Evaluation values of two fuel consumption models

	MSE	MAE	R_2
Random forest	0.0016	0.171	0.9969
LSTM neural network	0.0026	0.1742	0.9953

$$\mathrm{MAE} = \frac{1}{m} \sum_{i=1}^{m} \left| \hat{y}_i - y_i \right| \tag{20.6}$$

$$R_2 = \frac{\left(n \sum_{i=1}^{n} y_i \hat{y}_i - \sum_{i=1}^{n} y_i \sum_{i=1}^{n} \hat{y}_i \right)^2}{\left[n \sum_{i=1}^{n} {y_i}^2 - \left(\sum_{i=1}^{n} y_i \right)^2 \right] \left[n \sum_{i=1}^{n} y_i^2 - \left(\sum_{i=1}^{n} y_i \right)^2 \right]} \tag{20.7}$$

where $\hat{y}_i$ represents the predicted value of fuel consumption in the test sample, y_i is the true value of fuel consumption in the test sample, and m is the total number of samples. The mean square error (MSE), mean absolute error (MAE), and R_2 coefficients of the two fuel consumption models are shown in Table 20.2.

20.5 Conclusion

In this paper, the fuel consumption and related operating parameters collected by the T-BOX installed on the heavy commercial vehicle are collected, and the data is cleaned and standardized, and the random forest algorithm is used to build the fuel consumption prediction model of the heavy commercial vehicle, and the results are analyzed. Validation and comparative analysis. The model established in this paper can provide a certain theoretical basis for the subsequent research on fuel consumption prediction of heavy commercial vehicles. According to this study, the following conclusions can be drawn:

The MSE of the random forest fuel consumption prediction model based on the Python language is 0.0016, the MAE is 0.171, and the R_2 is 0.9967; the LSTM neural network model is 0.0026, the MSE is 0.174, and the R_2 is 0.9953. It can be seen that the random forest fuel consumption prediction model proposed in this paper has higher prediction accuracy than the LSTM neural network fuel consumption prediction model.

References

1. He, H., Jiangwei, C., Xifeng, A., Hong, L.: Construction of a Python-based vehicle fuel consumption prediction model [J]. Electron. Meas. Technol. **44**(20), 113–118 (2021)
2. Yalian, Y., Qiyuan, D., Qiangshou, L., Huanxin, P.: Construction of a neural network fuel consumption model for hybrid power coupling system [J]. J. Chongqing Univ. **42**(07), 1–9 (2019)
3. Xiaohui, S., Zhang Yuxi, X., Shuping, S.Y.: Research on driving conditions and fuel consumption of improved K-means clustering algorithm [J]. Comput. Eng. Sci. **43**(11), 2020–2026 (2021)
4. Wickramanayake, S., Bandara, D.: Fuel consumption prediction of fleet vehicles using machine learning: a comparative study [C]. In: IEEE 2016 Moratuwa Engineering Research Conference (MERCon), Moratuwa, Sri Lanka, 2016, pp. 90–95.
5. Jakov, T., Branimir, Š., Joško, D.: Neural network-based prediction of vehicle fuel consumption based on driving cycle data [J]. Sustainability. **14**(2), 73–74 (2022)
6. Capraz, A.G., Ozel, P., Sevkli, M., et al.: Fuel consumption models applied to automobiles using real time data: a comparison of statistical models [J], vol. 83, pp. 774–781. Procedia Comput. Sci. (2016)
7. Lili, Y., Weize, T., Yuanyuan, X., Caicong, W.: A random forest prediction model for oil consumption of grain combine harvesters [J]. Chin. J. Agric. Eng. **37**(09), 275–281 (2021)

Chapter 21
Highway Traffic Carbon Emission Estimation Based on Big Data of Electronic Toll Collection: A Case Study from Guangzhou, China

JunDa Huang, PengPeng Xu, HuiYing Wen, and Sheng Zhao

21.1 Introduction

The Signing of ⟨⟨The Paris Agreement⟩⟩ showed that many countries have already demonstrated their ambition to reduce carbon emissions [1, 2]. For example, in the European Union, 78% of cities have already set greenhouse gas reduction targets. Both Britain and China have pushed electric cars into their transport networks and are preparing to phase out the sale of gas-powered vehicles [3]. However, achieving the goal of carbon emission reduction and carbon neutrality is difficult and involves the cooperative efforts of various stakeholders. And it is well known that transportation is a key sector of greenhouse gas emissions [4]. Taking China as an example, in 2020 the transportation sector emitted 930 million tons of carbon emissions, accounting for 15% of the total carbon emissions. In the whole field of transportation, road transportation accounts for 90% of carbon emissions, of which, road passenger transport accounts for 42%, 90% of which come from passenger vehicles. Road freight accounts for 45% of emissions, mainly from freight trucks [5].

Currently, Vehicles driving in the road network to carry passengers and goods are dominated by fuel cars, which makes the carbon emissions of road transportation remain high, causing severe environmental problems.

J. Huang · P. Xu · H. Wen (✉) · S. Zhao
School of Civil Engineering and Transportation, South China University of Technology, Guangzhou, china
e-mail: cthuangjunda@mail.scut.edu.cn; peng90@scut.edu.cn; hywen@scut.edu.cn; ctszhao@scut.edu.cn

X. Wang (ed.), *Future Energy*, Green Energy and Technology,
https://doi.org/10.1007/978-3-031-33906-6_21

21.2 Literature Review

From the meso and micro perspectives, traffic volume, vehicle type composition, vehicle speed and other factors have a close relationship with carbon emission. Various vehicle carbon emission models have been established, such as the Mobile Source Emission Factor Model [6], Comprehensive Modal Emissions Model [7], Motor Vehicle Emission Simulator [8], and Computer Programme to Calculate Emissions from Road Transport [9].

From the macro perspective, the "top-down method" and "bottom-up method" provided by the Intergovernmental Panel on Climate Change (IPCC) are currently two well recognized and widely used methods for calculating carbon emissions from road or regional transportation [10].

When it is easy to obtain the fuel consumption data of various vehicle types, the "top-down method" is suitable. By multiplying fuel consumption by the average fuel emission factor, the emissions of vehicles on the road or within a region can be calculated. Jeng used big data of ETC from Taiwan National Highway Department to evaluate the fuel consumption and carbon emissions of five vehicle categories (i.e., cars, light-duty trucks, public transits, and heavy-trucks) by per ton/passenger kilometer, and the carbon emission distributions for various parts of the highway [11]. Jiang used "top-down" method to measure the traffic carbon emissions from 1985 to 2016 in the Yangtze River Economic Belt and analyzed its spatial pattern and temporal evolution characteristics [12].

When it is more convenient to obtain data such as the number of vehicles with different emission standards and the average mileage of vehicles, the "bottom-up method" is more suitable. By calculating the number of various models, the average mileage and the product of corresponding emission factors, the carbon emissions of road vehicles or regional vehicles can be calculated.Adhi used top-down and bottom-up carbon emission calculation methods respectively to evaluate the carbon emission of bus rapid transit and highway in Jakarta, Indonesia. The results proved that bottom-up method was more accurate when vehicle mileage data (VMT) was more accurate [13]. Xu calculated the CO2 emissions from 37 vehicular types and three fuel categories across 339 cities in China through the bottom-up method, and developed a national vehicular CO2 emission inventory in a high spatial resolution [14].

21.3 Data and Mathodology

As can be seen from the literature review, traditional studies mostly estimated the carbon emissions of the entire transportation industry or a large range of regions, while more detailed carbon emissions studies of roads, especially highways, are limited. The analysis and positioning of key emission road sections are even rare. Therefore, this paper attempts to make full use of the big data of highway ETC

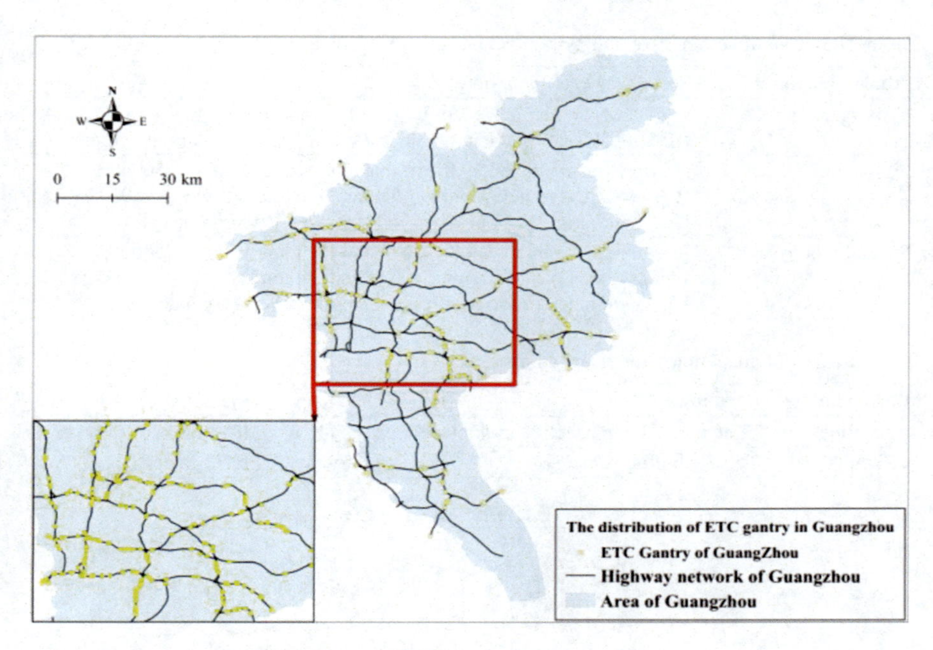

Fig. 21.1 Spatialization of ETC gantry in GIS

to establish a high-precision carbon emission measurement model to promote the carbon emission management of highway area.

After the cancellation of provincial border ETC stations in 2020, Guangzhou has also installed ETC gantries sufficient to cover the highway network. At present, there are 10 highways in Guangzhou with a total length of over 1000 kilometers, and more than 700 ETC gantries have been installed on highway within the city. The ETC system backstage generates more than 15 million pieces of data every day. The location of ETC gantries is illustrated (see Fig. 21.1).

In this study, the data collected by highway ETC system from November 1 to 30, 2021 were selected. Guangzhou was less affected by the coronavirus in this month, which can reflect the general pattern.

According to 《Reply of Guangdong Provincial People's Government on adjusting toll charging method of toll road》, Vehicles are divided into three categories: passenger car, freight car and special motor vehicle, as shown in Table 21.1.

The initial ETC big data contains 184 fields, but only a few fields need to be extracted to calculate the traffic volume of different types of vehicles, as shown in Table 21.2. After data cleaning, the daily flow data of different vehicle types on Guangzhou highway can be obtained (see Fig. 21.2). It can be found that the operating vehicles of Guangzhou highway were mainly passenger vehicles, accounting for almost 80%, followed by freight vehicles, accounting for almost 20%, and special vehicles account for the least, with the proportion of less than

Table 21.1 Classification of vehicle types in Guangzhou ETC data

Primary classification	Secondary classification
Passenger car	I: ≤9 seats and length < 6 m II: 10 ~ 19 seats and length < 6 m III: 20 ~ 39 seats and length ≥ 6 m IV: ≥40 seats and length ≥ 6 m
Freight car	I: Two-axles (vehicle length < 6 m and the total weight <4500 kg) II: Two-axles (vehicle length ≥ 6 m and the total weight ≥ 4500 kg) III: Three-axles IV: Four-axles V: Five-axles VI: Six-axles
Special motor vehicle	I: Two-axles (vehicle length < 6 m and the total weight <4500 kg) II: Two-axles (vehicle length ≥ 6 m and the total weight ≥ 4500 kg) III: Three-axles IV: Four-axles V: Five-axles VI: ≥six-axles

Table 21.2 Main data fields and its meaning

Field name	Meaning	Field name	Meaning
pass_time	Time when the transaction took place	pass_id	Identification number
vehicle_type	Type of billing vehicle	gantry_hex	ETC gantry ID

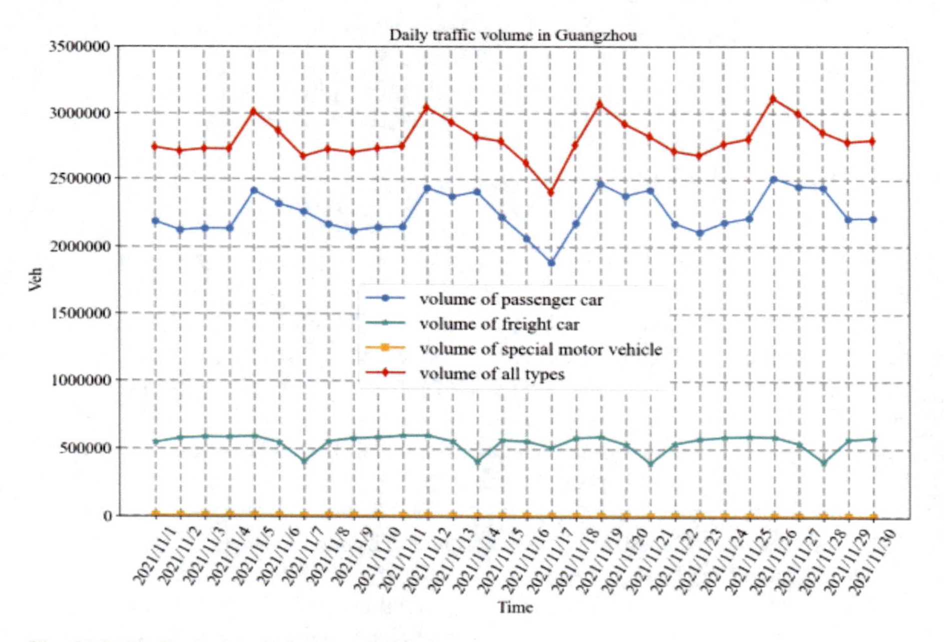

Fig. 21.2 Traffic flow in Guangzhou in November 2021

0.3%. Therefore, to avoid the inconvenience of calculation caused by too many vehicle types, we omitted the carbon emissions caused by special motor vehicles.

There are some problems in the traditional highway carbon emission calculation method, such as:

(1) It is often difficult to assess the road segment distribution of transport carbon emissions in order to achieve precise emission reduction;

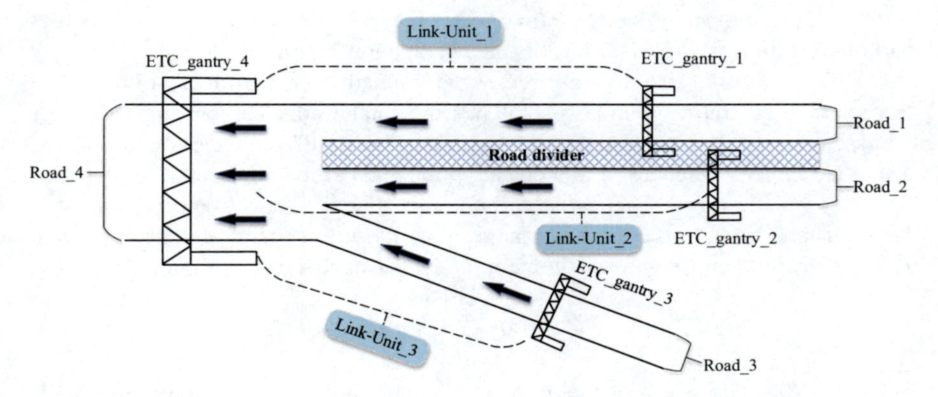

Fig. 21.3 How to define "Link-Unit" (BSU)

(2) It is also difficult to distinguish the carbon emission values of different types of vehicles for classified management.

In order to solve aforementioned problems, carbon emissions should be calculated at a more precise level (see Fig. 21.3), the trajectory of a vehicle can be traced based on the road profile and ETC data. The section between any two adjacent gantries with the same vehicle travel direction is called "Link-Unit" (i.e., basic spatial units, BSU). This also means that there may be multiple BSU on the same road segment.

Accordingly, excluding some ramps with short length, there are 1076 BSU after complete division. The calculation of traffic carbon emissions of subsequent highway will be based on them.

At present, the internationally recognized accounting method of carbon emissions is the method proposed by the Intergovernmental Panel on Climate Change in ⟨⟨IPCC 2006 Guidelines for National Greenhouse Gas Inventories⟩⟩, as shown in Eqs. (21.1) to (21.3).

$$CEF = NCV \times PF \times COF \times \theta \tag{21.1}$$

$$CEFL_k = CEF \times \rho_i \tag{21.2}$$

$$TCE_k = FC_k \times CEFL_k \tag{21.3}$$

Where TCE_k (kg) is the carbon emission of class K fuel; FC_k (L) is the consumption of class K fuel; $CEFL_k$ ($CO_2 \cdot kg^{-1}$) is the carbon emission coefficient of class K fuel; *CEF* ($CO2 \cdot kg^{-1}$) is carbon dioxide emission coefficient; *PF* (t-C/TJ) is the potential carbon emission coefficient, expressed by the mass of carbon element contained in fuel per unit calorific value; θ is the carbon conversion efficiency, that is, the relative molecular weight ratio of carbon dioxide, $\theta = 3.67$; COF (%) is fuel carbon oxidation rate; NCF (kJ/kg) is the average low calorific value of fuel.

Traffic carbon emissions are directly determined by road traffic volume. The larger the road traffic volume, the higher the total traffic carbon emissions. Of course, vehicle types are also another important factor affecting carbon emissions. There are large differences in fuel consumption and carbon emission rates between different vehicle types. The traffic carbon emissions in sections and regions with large bus and truck traffic are much larger than those in general regions. So, given BSU, the carbon emission measurement model is established as follows:

$$TCE_{i,j} = FC_j/100 \times LU_i \times Q_{i,j} \times CEFL_k \tag{21.4}$$

Where $TCE_{i,j}$ ($CO_2 \cdot kg^{-1}$) is the traffic carbon emission of Class j vehicles on BSU i; FC_j ($L \cdot 100\ km^{-1}$) is the fuel consumption of class j vehicles (L/100KM); LU_i (km) is the length of the BSU i; $Q_{i,j}$ (veh) is the traffic flow of class j vehicles on BSU i;

Based on all the above models, it can be concluded that the carbon emission calculation model of all vehicles on any road section is:

$$TCE_i = \sum_{j=1}^{10} TCE_{i,j} \tag{21.5}$$

Then, the calculation model of total carbon emission on all sections of the entire road network, including all types of vehicles, is as follows:

$$TCE = \sum_{i=1}^{n} TCE_i = \sum_{i=1}^{n} \sum_{j}^{10} TCE_{i,j} = \sum_{i=1}^{n} \sum_{j}^{10} FC_j/100 \times LU_i \times Q_{i,j} \times CEFL_k \tag{21.6}$$

$$= \sum_{i=1}^{n} \sum_{j}^{10} FC_j/100 \times LU_i \times Q_{i,j} \times NCV \times PF \times COF \times \theta \times \rho_i \tag{21.7}$$

Where TCE(kg) is total carbon emissions; n is the number of road sections in the study area.

Then refer to the data in ⟪General Principles for the Calculation of Comprehensive Energy Consumption: GB/T 2589-2008⟫ and ⟪China Energy Statistical Yearbook 2018⟫, the values of each parameter in the formula are shown in Table 21.3.

Based on Eq. (21.1), it can be calculated that the CEF values of carbon dioxide emission factors of diesel and gasoline on highway in China were 3.09 and 2.93, respectively, that is, 3.09 kg or 2.93 kg carbon dioxide gas was expected to be produced for every 1 L diesel or gasoline consumed by vehicles.

Table 21.3 China highway gasoline and diesel energy information table

Fuel type	Density(kg·L^{-1})	Average lower calorific value(kJ·kg^{-1})	Potential carbon emissions factor(tC·TJ^{-1})	Carbon oxidation rate(%)
Diesel	0.84	42,652	20.17	98
Gasoline	0.75	43,070	18.90	98

Table 21.4 fuel consumption of different models in Guangzhou(L·100 km^{-1})

Primary classification	Specific type	Fuel consumption (L·100 km^{-1})	Primary classification	Specific type	Fuel consumption (L·100 km^{-1})
Passenger car	I	7.3	Freight car	I	14.4
	II	13.1	II	20.1	
	III	19.7	III	24.8	
	IV	24.5	IV	31.2	
–	–	–	V	33.6	
–	–	–	VI	41.3	

According to the big data of car ownership in Guangzhou, in 2020, the number of motor vehicles in Guangzhou was about 3,081,800, including 2,527,600 passenger cars and 435,800 freight cars, of which 84.3% were gasoline consuming vehicles in class I passenger cars and 87.1% were diesel consuming vehicles in class II ~ IV passenger cars. The proportion of vehicles with diesel fuel as the energy supply mode for class I-VI trucks is 98.8%. According to the "automobile energy consumption query platform of the Ministry of industry and information technology of China", the average fuel consumption per 100 kilometers of 10 specific models in Guangzhou can be obtained, as shown in Table 21.4.

21.4 Case Study

By collecting and calculating the daily traffic flow data recorded by the highway ETC system in November 2021, it can be known that the average daily carbon emission of Guangzhou highway was 12,754 tons, among which the average daily carbon emission of freight car was 6981.9 tons, accounting for 54.7% of the total carbon emission. The carbon emission of passenger cars was 5772.1 tons, accounting for 45.3%. It can be seen that the main source of carbon emission of Guangzhou highway was freight car. Furthermore, different types of vehicles were analyzed (see Fig. 21.4).

Specifically, carbon emissions mainly resulted from small and medium-sized vehicles. For example, type I passenger car produced the largest carbon emissions, accounting for 93.0% of the total carbon emissions of passenger car and 42.1% of the total carbon emissions. While the most carbon emission in freight car was type

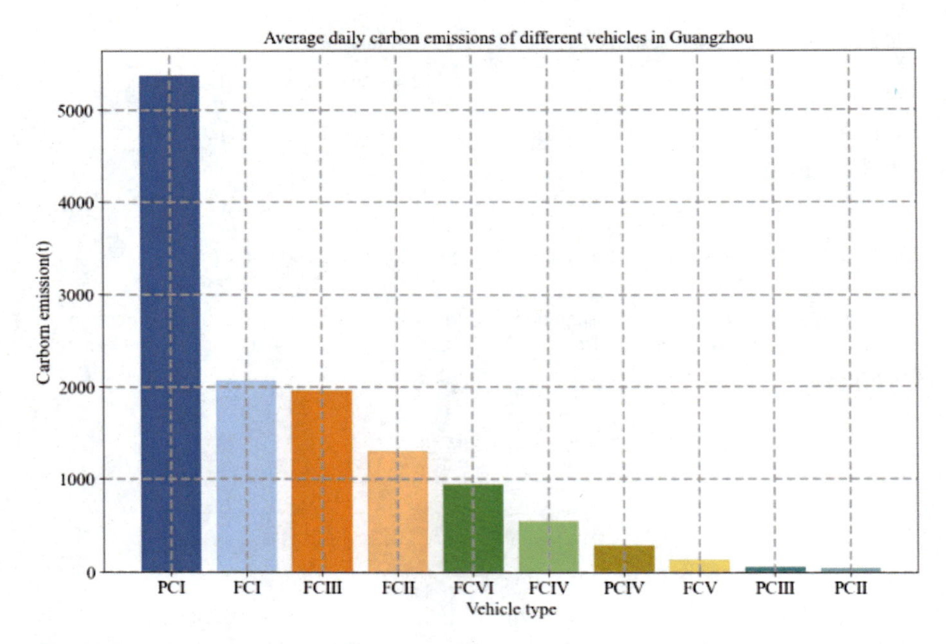

Fig. 21.4 Average daily carbon emissions of different vehicles in Guangzhou

I and type III, with a carbon emission of 2070.4 tons and 1963.7 tons, respectively, accounting for 57.8% of the total carbon emission of freight car.

In addition, in passenger car, except for type I, the general emission was low, accounting for only 7%. Among them, the carbon emission value of type II cars was the lowest (46.4 tons), accounting for only 0.80% of the total carbon emission of passenger cars, followed by the type III passenger cars (62.9 tons), accounting for 1.1% of the total carbon emission of passenger cars.

Since the defined BSU overlap on the same road section, in order to more clearly show the difference in the intensity of traffic carbon emissions on different BSU, we divided the whole highway network into regular grids with a size of 1 km*1 km, Then, the carbon emission on the BSU was converted into the carbon emission value per kilometer unit, and its average value was calculated in each grid, which was defined as the carbon emission intensity and visualized in ArcGIS. The carbon emission intensity of the road section is divided into 10 levels (see Fig. 21.5).

It can be seen that the spatial distribution of carbon emissions from highway in Guangzhou has the following significant characteristics:

(1) the carbon emission regions of the highway show obvious variations. In the context of the close connection among cities in the Greater Bay Area, the carbon emission level of the main corridors connecting Guangzhou with Foshan, Qingyuan, Huizhou, Dongguan and other cities is significantly higher.

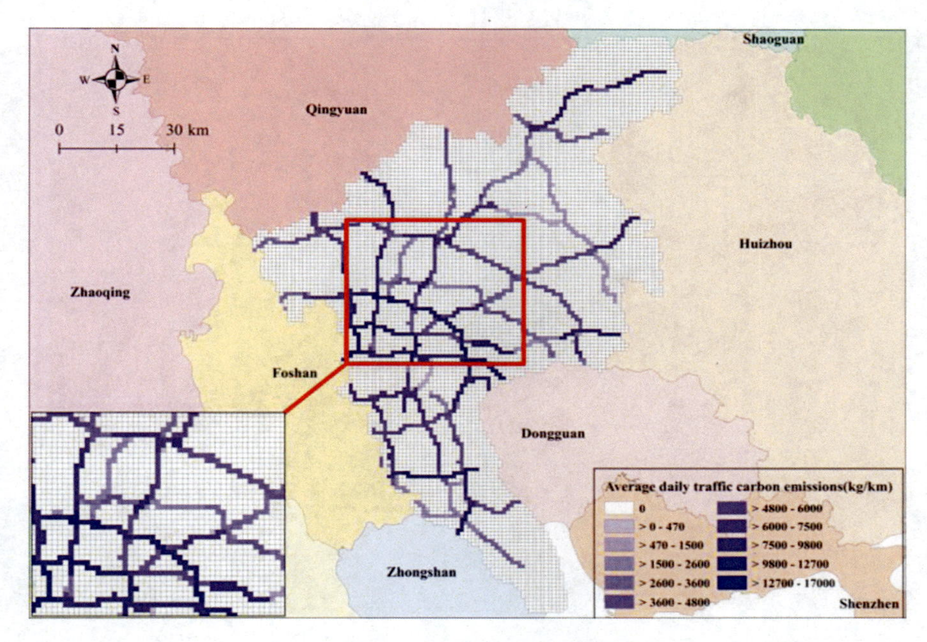

Fig. 21.5 Spatial distribution pattern of daily traffic carbon emissions in Guangzhou highway network

(2) in the core area of Guangzhou, especially in the downtown area of Huan-Cheng highway and Rao-Cheng highway, carbon emissions are also very high.
(3) Due to the large flow of highway in Guangzhou and the inflow of vehicles from multiple directions, At the junction of many Interchanges, the carbon emission level is very high.

In addition to obtaining the carbon emission distribution of the whole road network, it can also obtain the daily average traffic carbon emissions in all directions of the interchange for a special area, such as the North interchange airport (see Fig. 21.6), so as to achieve more refined management.

21.5 Conclusions and Recommendations

The model and analysis method established in this paper can estimate the carbon emission and its spatial distribution pattern of highway area more intuitive and detailed.

On the one hand, it can be clearly seen from the calculation results of traffic carbon emissions on Guangzhou highway that freight cars are the main source of traffic carbon emissions, accounting for 54.7%. The specific carbon emissions of freight cars are type "FCI", type "FCIII", and type "FCII" in order. The type

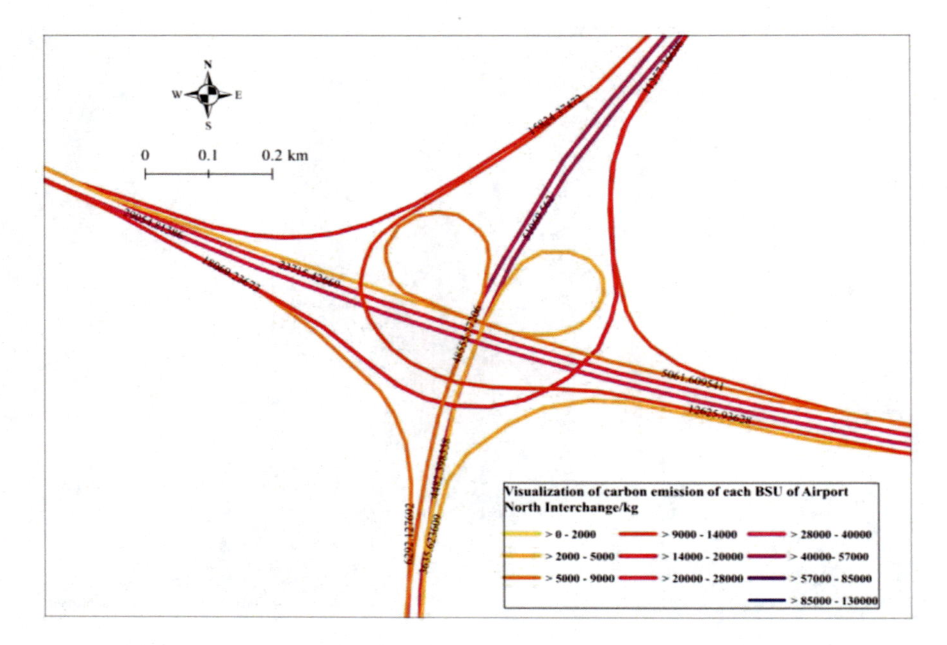

Fig. 21.6 Visualization of carbon emission of each BSU of Airport North Interchange

"PCI" accounted for a relatively high proportion of carbon emissions, reaching an astonishing 93.0%.

On the other hand, as a busy area in South China, the main connecting corridors between Guangzhou and other neighboring cities show a high carbon emission level, as well as the highway interchanges with heavy traffic in the city.

Therefore, in view of such a situation, some targeted measures can be put forward, such as: 1) focus on monitoring the main two types of freight cars (type "FCI" and "FCII"), avoid their frequent empty load, and can also consider guiding their driving path, 2) Focus on improving the energy structure of type "PCI" and increasing the penetration rate of new energy vehicles, 3) Accelerate the construction of important corridors in Guangzhou to solve the problem of high carbon emissions caused by serious congestion in some Interchanges, 4) focus on improving the overall composition of the overall vehicle structure, eliminate old vehicles, and use stricter emission standards to improve the regional carbon emission reduction capacity.

Based on the big data of ETC and the spatial characteristics of the ETC gantry position, this study established a high-precision measurement method of highway traffic carbon emissions. However, this paper does not consider the impact of terrain, road alignment, weather, speed and other factors on carbon emissions, which may be a promising direction for future research.

Acknowledgement The authors would like to thank the anonymous reviewers for their valuable comments and this study is funded by research grants from Fundamental Research Funds for the Central Universities (No. D2220800) and the National Natural Science Foundation of China (No. 52172345).

References

1. Salvia, M., Reckien, D., Pietrapertosa, F., et al.: Will climate mitigation ambitions lead to carbon neutrality? An analysis of the local-level plans of 327 cities in the EU. Renew. Sust. Energ. Rev. **135**, 110–253 (2021)
2. Logan, K.G., Nelson, J.D., Lu, X., et al.: UK and China: will electric vehicle integration meet Paris agreement targets? Transportation Research Interdisciplinary Perspectives. **8**, 100–245 (2020)
3. Yaacob, N.F.F., Mat Yazid, M.R., Abdul Maulud, K.N., et al.: A review of the measurement method, analysis and implementation policy of carbon dioxide emission from transportation. Sustainability. **12**, 58–73 (2020)
4. Song, D.Y., Song, X.Y., Zhang, L.: An analysis of the driving factors of China's transportation carbon emissions: based on decoupling theory and generalized fisher index decomposition. Science and Technology Management Research. **42**, 216–228 (2022)
5. McDonald, B.C., McKeen, S.A., Cui, Y.Y., et al.: Modeling ozone in the eastern US using a fuel-based mobile source emissions inventory. Environ. Sci. Technol. **52**, 7360–7370 (2022)
6. Zhu, L., Hu, D.: Study on the vehicle routing problem considering congestion and emission factors. Int. J. Prod. Res. **57**, 6115–6129 (2022)
7. Campbell, P., Zhang, Y., Yan, F., et al.: Impacts of transportation sector emissions on future US air quality in a changing climate. Part I: projected emissions, simulation design, and model evaluation. Environ. Pollut. **238**, 903–917 (2018)
8. Lejri, D., Can, A., Schiper, N., et al.: Accounting for traffic speed dynamics when calculating COPERT and PHEM pollutant emissions at the urban scale. Transp. Res. Part D: Transp. Environ. **63**, 588–603 (2018)
9. Eggleston, H. S., Buendia, L., Miwa, K., et al.: IPCC guidelines for national greenhouse gas inventories. 2006
10. Jeng, S.L., Chang, C.C., Lin, W.C., et al.: Using big data of electronic toll collection to analyze the energy consumption and carbon emissions of highway vehicles in Taiwan-a case study. Journal of the Chinese Statistical Association. **59**, 145–171 (2021)
11. Jiang, Z.R., Jin, H.H., Wang, C.J., et al.: Measurement of traffic carbon emissions and pattern of efficiency in the Yangtze River Economic Belt (1985–2016). Huan Jing ke Xue=Huanjing Kexue. **41**, 2972–2980 (2020)
12. Adhi, R.P.: Top-down and bottom-up method on measuring CO2 emission from road-based transportation system (case study: entire fuel consumption, bus rapid transit, and highway in Jakarta, Indonesia). Jurnal Teknologi Lingkungan. **19**, 249–258 (2018)
13. Xu, Y., Liu, Z., Xue, W., et al.: Identification of on-road vehicle CO2 emission pattern in China: a study based on a high-resolution emission inventory. Resour. Conserv. Recycl. **175** (2021)
14. Hu, Y. C., Wu, H., Huang, J. X., et.al. A method for estimating expressway section average speed based on support vector regression. Journal of Highway and Transportation Research and Development., vol. 36, pp. 137-143, 2019

Chapter 22
Carbon Capture and Storage (CCS) for India: Bottlenecks and Their Role in Adoption

T. Joji Rao and Krishan Kumar Pandey

22.1 Introduction

The worldwide emissions are ever increasing despite adoption of newer sources and energy efficient equipment. This establishes that adoption of renewable energy will not be adequate to reduce CO_2 to restrict the unfavourable influences of climate change. The targets to curb growing atmospheric CO_2 presence whilst meeting ever increasing energy demand can be attained by employing a thorough assortment of technologies that ought to consist of carbon capture, utilization, and storage (CCS) [1]. CCS includes approaches and technologies to take out CO_2 from the flue and from atmosphere, followed by either storage or recycling the CO_2 for utilization. CCUS is a crucial emissions lessening technology that may be employed in the industries. CCUS may also offer the basis for carbon removal which in current popular terms called as "negative emissions".

22.2 India's Intended Nationally Determined Contribution (INDC) in Energy Perspective

Intergovernmental Panel on Climate Change (IPCC) is striving to keep the global mean temperature below 2 °C by 2050. Countries like Japan, USA, China and EU have potential to achieve much more what they have their Intended Nationally Determined Contribution (INDC) targets [2]. For example, the approach that can be applied is to meet the INDC goals of Caribbean and South America efficiently is

T. J. Rao (✉) · K. K. Pandey
Jindal Global Business School, O P Jindal Global University, Sonipat, India
e-mail: tjrao@jgu.edu.in; kkpandey@jgu.edu.in

X. Wang (ed.), *Future Energy*, Green Energy and Technology,
https://doi.org/10.1007/978-3-031-33906-6_22

decrease consumption of crude oil. But in Central America strict energy efficiency policies needs to be implemented to maintain a balance between economic growth and preservation of crude oil. Policies aimed at achieving INDC targets have been unsuccessful in nearly all major economies [3]. Even China cannot reach its INDC targets without expanding the stake of biomass from 0.18% in 2010 to 10.20% and 20.06% in 2030 and 2050 respectively [4]. However, the Indian government is stepping forward with ways to build a low carbon intensity economy. For instance, India has lately announced a market-based PAT (Perform, Achieve and Trade) scheme. It has upscaled its renewable energy goal to 40 percent by 2030 and grew their electric vehicles (EVs) target by 2030. India has recently announced to switch all government vehicles to EVs. But these measures are not enough. More interventions are required from from government to achieve low carbon economy. As stated by the IPCC, we cannot achieve climate objectives without Carbon Capture and Storage (CCS). Therefore, to take the first step toward CCS in India a detailed study needs to be conducted on feasibility of CCUS technology in India. From the perspective of technological feasibility few research has quantified the potential of carbon capture in India and identified the geographical mapping of the potential [5–12]. But a prior understanding of the major challenges and the interrelationship among them needs to be the first step before going for detailed feasibility study.

22.3 Objective of the Study

I. To identify the major challenges of carbon capture and storage adoption in India from the perspective of emitter organizations.
II. To Study the interrelationship among the identified challenges.

The detailed methodology and findings are discussed as follows:

22.4 Major Challenges of CCSU Adoption (Objective I)

To achieve the objective of identifying the major challenges of CCSU adoption, participatory approach was adopted. The common elements of participatory approaches are (a) Common Elements (b) Focuses on being useful (c) Employs diverse methods (d) Emphasizes collaboration [13]. The precise research tool used for participatory approach in the study is focused group.

The study accepted focus groups because they are an especially useful way to recognize people's ideas and opinions about a particular topic. The focused group was consisted of 10 participants ensuring equal representation from upstream oil and gas companies which are into midstream as well and fossil fuel-based power generation companies in India. The participants were at the senior strategic management level with a considerable experience in running the plant operations. The focused group was moderated with open ended questions intending to understand the challenges of carbon capture and storage initiatives in Indian context.

22.4.1 Findings

The carbon capture and storage is at a very nascent stage in India. The industry is aware of the technology and its significance. The respondents seemed to be very sceptical about the way it is being viewed in India currently. As per the response the major challenges of concern while introducing the Carbon capture and storage initiatives may be broadly put into 6 categories namely (a) Cost of CCS (b) Geo-storage capacity (c) Source sink matching (d) Supply Chain and building rate (e) Policy regulations and marketing (f) Public Acceptance.

Cost of CCS. Cost of CCS

It has been found as the major challenge inhibiting the widespread adoption of this technology. The study found estimating actual CCS cost and expressing it in a clear way is challenging. This is due to the lack of data, difficulty in choosing the baseline when comparing different CCS plants, cost differences due to unavailability of transport and storage infrastructure and a variety of processes, operating conditions, and capture processes.

Geo-Storage Capacity

The respondents suggest that there is an enormous capacity in depleted oil and gas reservoirs. Hence, they have not recommended the capacity as a challenge. But they feel that in saline aquifers (fields devoid of hydrocarbon production) reservoir pressurisation will restrict the accessible CO_2 geo-storage capacity. The study also suggests that potential limit on storage capacity on the deployment of CCS in integrated assessment models may be a challenge.

Source-Sink Matching

Source-sink matching shows that CCS will not be constrained by local availability of storage resources. Outside of these areas, storage availability is highly uncertain. Hence there may be few locations where availability of local storage may be a limiting factor.

Supply Chain and Building Rate

In Indian context the supply chain issues has been identified as a major challenge for CCS initiatives. The potential supply chain issues may be availability of hydrogen turbines for the capture step, availability of pipelines for the transport step, availability of geo-engineers and drilling rigs for the storage step and availability of petroleum engineers across the full CCS chain.

Policy, Regulations, and Market

The findings of the study suggest that India is in need of a dynamic policy regulations and market. The findings recommend possible policy options for carbon trading mechanism. The industry feels that the government needs to place a very robust carbon taxation mechanism before promoting carbon capture and storage initiatives. The policy needs to promote investment support for the initial capital costs; feed-in schemes and a favourable carbon floor price.

Public Acceptance

Public acceptance has a key role in the deployment of carbon capture and storage, locally and globally. The identified public acceptance constructs of the study are attitude, knowledge, experience, trust, fairness, perceived costs, risk and benefits, outcome efficacy and problem perception.

22.5 Interrelationship among Identified Challenges of CCSU Adoption (Objective II)

After identifying and extracting the challenges and categorizing them into 6 categories a hybrid MICMAC-ISM method has been employed to categorise the factors and present a conceptual model to establish the relationship between them. When numerical information is not available and the factors are not quantifiable or qualitative in nature, application of an expert-based decision-making techniques e.g., ISM, Decision making trial and evaluation laboratory (DEMATEL), to represent the relationship between the factors is suggested [14]. Even as DEMATEL tool can analyse the trigger-and-impact relationship between the factors, the ISM-MICMAC hybrid methodology categorises the factors into four clusters and shows a cause-effect illustration and a stage-based conceptual model. Moreover, ISM algorithm verifies the validity of the questionnaires finished by the experts and prevents inconsistency.

22.5.1 *Findings*

Final Reachability Matrix

The Initial matrix attained from the data analysis is transformed to zero and one matrix (ZOM. In this stage, the validity of the expert's opinions is checked and for possible inconsistencies, revision is performed on the ZOM to build a reliable matrix for further calculations which is called FRM. Reachability matrix is derived from SSIM using the following rules:

If Xi,j entry in SSIM is V, then Xi,j is set to 1, and Xj,i is set as 0;

(i). If Xi,j entry in SSIM is A, then Xi,j is set to 0, and Xj,i is set as 1;
(ii). If Xi,j entry in SSIM is X, both Xi,j and Xj,i is set as 1;
(iii). If Xi,j entry in SSIM is O, both Xi,j and Xj,i is set as 0.

The final reachability matrix of the challenges of CCSU adoption was developed which resulted into a matrix for 22 variables under study. The reachability matrix was further converted into Level partitioning as shown below:

Level Partitioning (LP)

The reachability matrix found above is further partitioned into levels. The reachability and antecedent sets for each category were originate from the values in Table 2. The reachability set for an individual category consists of itself and the other categories which may have effect on. The antecedent set of an individual category is the list of the categories themselves which may have effect on it. The intersection of these sets was also derived for all categories.

Relational Diagram (MIMAC Analysis)

The objective of the MICMAC analysis is to evaluate the driving power and the dependence power of variables.

The rational diagram of the identified challenges is shown in Fig. 22.1.

According to the answers from the experts, all challenges except availability of expert geological engineers and petroleum engineers fell into this Classification. This means that these categories are unstable in the fact that any action against the barriers of these categories will affect others and feedback on themselves. Furthermore, as many categories are in this sector, the system is strongly connected and unstable. Only the challenge of sourcing expert geological engineers and petroleum engineers Classification fell into the fourth sector, which includes the independent categories having strong driving power but weak dependence.

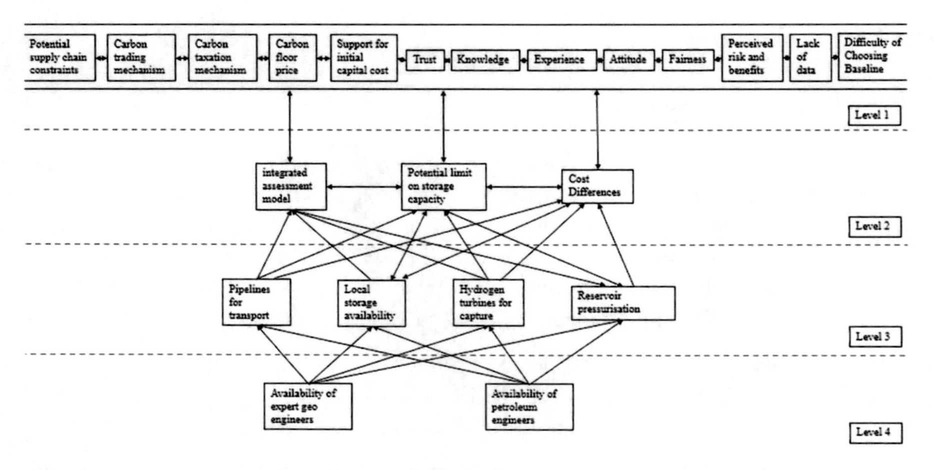

Fig. 22.1 The rational diagram of the identified challenges of CCS adoption

22.6 Discussion

Carbon Capture and Storage (CCS) is in a very infant stage in India. It is expected that it will gain momentum in terms of sizeable adoption in next 5 to 8 years. The study identified 22 challenges categorised into 6 categories. There is a need to adopt a proactive approach to ensure regulations, tools, models, support systems at place to witness successful CCS adoption unfold in coming years. The relationship of these identified challenges establishes the following major findings:

A. A presence of Carbon trading and Carbon taxation mechanism will lead to trust, knowledge, experience, attitude, and fairness.
B. Perceived risks and benefits, lack of data and difficulty in choosing baseline will impact the trust, knowledge, experience, attitude, and fairness.
C. Integrated assessment model, cost difference and potential limit on storage are interlinked.
D. Hydrogen turbines for capture and reservoir pressurization will lead to cost differentiation.
E. Availability of expert geological and petroleum engineers will impact pipeline for transport and application of hydrogen turbines.

The most significant policy measures that may be inferred from the findings of the study are as follows:

A. India must ensure a vibrant carbon trading and carbon taxation mechanism at place in priority. It must be treated as a prerequisite not as a support mechanism for CCS adoption in India.

B. Adequate policy measures and initiatives must be ensured to make the players aware of the perceived risks, benefits, data sourcing and methods to reach to a baseline. It will ensure trust, attitude, and fairness among the CCS adopters.
C. Integrated assessment models need to be worked out before the policy measures and it will give directions to the regulations. It will clear the ambiguity about cost, base line and modus operandi of the process.

22.7 Research Limitations/Implications

The results reflected or obtained via this study are based on experts' opinions which sometimes are biased in nature and influence the final output of any empirical structural model. The research implications are basically made to help the industry to understand the critical challenges and to prioritize or eliminate the challenges in CCS sector.

22.8 Originality/Value

A First-time attempt is made to understand the very critical and imminent challenges of carbon capture and storage adoption in India. The ISM methodology is also applied for the first time for this type of work. This study pave a way for the various researcher in understanding the various barriers and their interrelationships. This paper will surely help in removing various barriers against CCS implementation in India.

22.9 Conclusion

The world can be clearly divided into two segments in terms of CCS as a tool of decarbonisation. Firstly, the developed countries who have identified the significance of CCS and are gradually adopting it despite high cost of capture and storage. The other segment of the world is the one which is yet to gain the confidence of accepting CCS as a solution to high greenhouse gas emissions. The study attempted to establish this gap by comparing India with rest of developed countries who are gradually accepting CCS. For India the challenges are different. The Cost, technology, storage, policy, and acceptance are major challenges. The policy makers need to prioritise these challenges in their efforts to create a conducive environment for the industry to accept CCS as a tool for their efforts towards decarbonisation.

References

1. Pacala, S., Socolow, R.: Stabilization wedges: solving the climate problem for the next 50 years with current technologies. Science. **305**(5686), 968–972 (2004)
2. Ke-Jun, J., Tamura, K., Hanaoka, T.: Can we go beyond INDCs: analysis of a future mitigation possibility in China, Japan, EU and the U.S. Adv. Clim. Chang. Res. **8**, 117–122 (2017)
3. Behmiri, Niaz Bashiri; Manso, Jose Ramos Pires; The linkage between crude oil consumption and economic growth in Latin America: the panel framework investigations for multiple regions; Energy, Vol 72, Pages 233–241,(2014)
4. Hui, J., Cai, W., Wang, C.: Achieving China's INDC: biomass development and competition for land. Energy Proceedia. **105**, 3521–3126 (2017)
5. Gupta, A., Paul, A.: Carbon capture and sequestration potential in India: a comprehensive review. Energy Procedia. **160**, 848–855 (2019)
6. Aankur Malyan and Vaibhav Chaturvedi; Carbon capture, utilisation, and storage (CCUS) in India from a cameo to supporting role in the Nation's low-carbon story, CEEW Council, 1, page 1 to 15, (2021)
7. R. Beck, Y. Price, S. Friedmann, L. Wilder and L. Neher, "Mapping highly cost-effective carbon capture and storage opportunities in India," J. Environ. Prot., Vol. 4 No. 10, 2013, pp. 1088–1098, (2013)
8. Omkar, S.: Patange, Amit Garg, Sachin Jayaswal, an integrated bottom-up optimization to investigate the role of BECCS in transitioning towards a net-zero energy system: a case study from Gujarat, India. Energy. **255**, 124508 (2022)
9. Mishra, G.K., Meena, R.K., Mitra, S., Saha, K., Dhakate, V.P., Prakash, O., Singh, R.K.: Planning India's first CO2-EOR project as carbon capture utilization & storage: a step towards sustainable growth. In: Paper Presented at the SPE Oil and Gas India Conference and Exhibition, One Petro, Mumbai, India (2019)
10. Viebahn, P., Vallentin, D., Höller, S.: Prospects of carbon capture and storage (CCS) in India's power sector – an integrated assessment. Appl. Energy. **117**, 62–75 (2014)
11. Shaw, R., Mukherjee, S.: The development of carbon capture and storage (CCS) in India: a critical review. Carbon Capture Science & Technology. **2**, 100036 (2022)
12. Harsha Kumar Bokka: Kai Zhang. Hon Chung Lau, Carbon capture and storage opportunities in the west coast of India, Energy Reports. **8**(2022), 3930–3947 (2022)
13. Randy, S.: Challenging institutional barriers to community-based research. Action Res. **6**(I), 49–67 (2008)
14. Amoozad Mahdiraji, H., Hafeez, K., Kord, H., Abbasi Kamardi, A.: Analysing the voice of customers by a hybrid fuzzy decision-making approach in a developing country's automotive market. Manag. Decis. **60**(2), 399–425 (2022)

Chapter 23
Quantitative Assessment of Low-Carbon Transition Pathways of Power Generation Company Considering CCS Technology

Zhou Yu, Xue Feng, Cai Bin, and Xue Yusheng

23.1 Introduction

CCS refers to a low-carbon technology that separates CO2 from industrial sources or stores it directly [1]. Because of its huge emission reduction potential and wide applicability, CCS technology is recognized as one of the most important technical means to achieve low-carbon development [2]. The International Energy Agency (IEA) believes that majority of climate models will not meet deep emission reduction targets without CCS [3]. CCS transition can reduce the CO2 emissions of coal power units to 10~20% of the original value, which can provide indispensable flexible low-carbon power for the safe and reliable operation of new power systems, and is one of the most important and promising CCS application fields [4–6].

Currently, the CO2 capture capacity of CCS facilities worldwide is about 40 million ton/year, compared to only three million ton/year in China. There are about 40 CCS demonstration projects in China that have been put into operation or are under construction, including the CO2 capture project of Guohua Jinjie Power Plant of China Energy Group (capture capacity of 150,000 t/year) and the CO2 separation and liquefaction project of China National Offshore Oil Corporation (CNOOC) Lishui gas field (capture capacity of 250,000 t/year). According to the 2021 China Carbon Dioxide Capture Utilization and Storage Annual Report, in order to achieve the goal of carbon neutrality, the annual capture capacity of CCS in China is expected to reach 2 billion tons. However, because CCS costs are still very expensive, in the absence of subsidies, CCS projects are not profitable enough and even have large losses [7].

Z. Yu (✉) · X. Feng · C. Bin · X. Yusheng
The NARI Group Corporation, Nanjing, China
e-mail: xue-feng@sgepri.sgcc.com.cn; caibin@sgepri.sgcc.com.cn; xueyusheng@sgepri.sgcc.com.cn

X. Wang (ed.), *Future Energy*, Green Energy and Technology,
https://doi.org/10.1007/978-3-031-33906-6_23

Power generation companies are the promoters and practitioners of carbon neutrality in energy and electricity. In addition to considering the requirements of power supply and low-carbon transition, the transition and development of power generation companies also need to ensure their economic benefits. At present, new energy power generation maintains a good return on investment. Developing new energy vigorously is the core development strategy and inevitable trend of power generation companies under the background of carbon neutrality. Therefore, it is necessary to introduce corresponding incentive policies to promote the coordinated development of new energy and coal power CCS by power generation companies. Assessing the impact of CCS on the low-carbon transition pathway and economic benefits of power generation companies quantitatively is the key to support relevant policy decisions.

The literature [8] proposes the concept of cyber-physical-social system in energy (CPSSE), which provides a research framework for the cross-domain energy transition problem integrating information and physical and social elements. Literature [9] proposes a hybrid simulation framework covering physical infrastructure of power, energy suppliers and external environments such as energy policy for CPSSE problems such as energy transition. Based on this framework, the technical-economic-emission simulation model developed on the Sim-CPSS platform [10] has been used in the energy transition research of nations [11], provinces [12], and companies [13].

The quantitative assessment of the low-carbon transition pathway of power generation companies considering CCS involves not only physical factors such as electricity and carbon emissions but also economic and social factors such as electricity prices, fuel prices, carbon prices, carbon quotas, and company capital flows. It is a typical CPSSE problem. Therefore, under the guidance of the CPSSE framework, this study will quantitatively analyze the impact of large-scale CCS development on the low-carbon transition of a power generation company by the technical-economic-emission simulation model based on the Sim-CPSS platform and provide decision-making support for the CCS development pathway strategy and related policies.

23.2 Research Method and Parameter Settings

23.2.1 Research Method

Based on the simulation model of technology-economy-emission of power generation companies transition considering CCS, this paper constructs a transition pathway for a power generation company without CCS (hereinafter referred to as "Without CCS pathway") and three transition pathways for that power generation company considering CCS (hereinafter referred to as "CCS development pathway") and then simulates them separately; extracts the indicators of electricity, emissions,

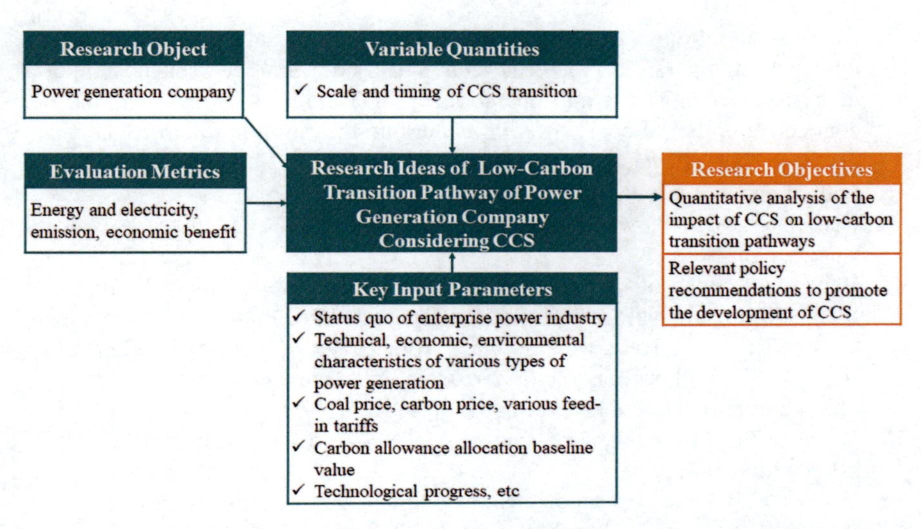

Fig. 23.1 Research method of low-carbon transition pathway of power generation company considering CCS

and economy from the simulation results; and quantitatively analyzes the specific impact of CCS transition on the low-carbon transition of that power generation company and related policy recommendations (see Fig. 23.1).

23.2.2 Scenario and Parameter Settings

Referring to relevant policy documents and industry norm values [14], some parameters of simulation deduction are set as follows:

1. Evaluation period: 2021 is the starting year, and 2021–2060 is the evaluation period.
2. Power generation types: covering coal power, hydropower, wind power (distinguishing between onshore and offshore), photovoltaic, nuclear power, etc.
3. Total annual power generation: In 2021, it will be 400 billion kWh; set every 5 years as a node year, the growth rate between node years will increase in equal ratio, of which the total power generation in 2025, 2030, 2050, and 2060 will be 485, 560, 760, and 800 billion kWh, respectively.
4. Power generation structure: In 2021, coal, onshore wind, offshore wind, photovoltaic, hydropower, and nuclear power accounted for 80.0%, 2.5%, 1.5%, 3.0%, 10.0%, and 3.0% of the power generation, respectively; in 2060, the above proportions will be 6.0%, 27.0%, 18.0%, 30.0%, 10.0%, and 9.0%, respectively.

5. Coal consumption of coal power units: The coal consumption of coal power units put into operation in 1991 is set at 340 g/kWh, and the coal consumption of coal power units put into operation in 2025 is set at 280 g/kWh, and the coal consumption of coal power units put into operation in intermediate years decreases linearly year by year.
6. Fuel price (coal price): 600 yuan/t standard coal.
7. Carbon emission allowance price: 60 yuan/t CO2 in 2021 and increases 6 yuan/t CO2 per year.
8. Benchmark value of carbon allowances for power supply: In 2021, referring to the "Implementation Plan for Setting and Allocation of Total National Carbon Emission Trading Allowances from 2019 to 2020 (Power Generation Industry)," it will decline to 0 by 2050 in a linear manner.
9. CO2 capture rate of coal power units: 90%.
10. Energy efficiency loss of CCS transition: decrease from 20% in 2031 to 15% in 2060 with equal ratio.
11. Unit CCS transition operation cost (excluding energy efficiency loss): decrease from 300 yuan/t in 2031 to 200 yuan/t in 2060 with equal ratio.
12. Service life of coal power units: 40 years for units without CCS and 15 years else for units with CCS.
13. Annual power generation hours of coal power units: the annual power generation hours of coal power in 2021 without CCS units will be 4000 h, which will drop linearly to 3200 h in 2060, and CCS units will add an additional 500 h/year.
14. Coal electricity price: It is extremely difficult to predict the trend of on-grid electricity price of each power generation type during the entire transition period, assuming that the profit of coal power kWh is maintained at 0.04 yuan/kWh, and the on-grid tariff is set accordingly.
15. New energy on-grid electricity price: Onshore wind power, offshore wind power, and photovoltaic feed-in tariffs in 2021 will be 0.40, 0.70, and 0.40 yuan/kWh, falling to 0.28, 0.35, and 0.24 yuan/kWh in 2060, with a decrease of 30%, 50%, and 40%, respectively.

It is worth pointing out that the parameter values affect the specific numerical results of simulation deduction, but do not affect the validity of the research method proposed in this paper.

23.2.3 CCS Transition Pathways

Combined with the operation and development characteristics of CCS transition of coal power units [15], the construction of *CCS development pathways* follows the following principles:

1. In the order of time urgency, priority is given to units that were retired earlier, that is, units that were put into operation earlier. Set up units that need to be renovated in the 35th year of operation.

Table 23.1 Typical CCS development pathway setting

Typical CCS development pathway	Number of units transformed per year	Total number of units transformed
Baseline CCS	2031–2045: 1 unit/year 2045–2060: 2 units/year	45 units(1/2 of total)
High CCS	2031–2040: 1 unit/year 2041–2050: 2 units/year 2051–2055: 3 units/year	60 units(2/3 of total)
Low CCS	2031–2060: 1 unit/year	30 units(1/3 of total)

2. The development pathway of CCS is divided into two stages: the industrialization incubation period before 2030 and the large-scale promotion period after 2030. This study regards the large-scale development of CCS as the research object and sets 2031 as the year when CCS transition is first implemented.
3. The basic unit of CCS transition is set to entire coal power unit, that is, 90% of the emissions of the entire coal power unit. The capacity of all coal power units of the target power generation company is 1000 MW, a total of 90 units.

According to the above principles, three typical *CCS development pathways* are designed as shown in the following table (Table 23.1).

23.3 Analysis of the Results

23.3.1 Electricity Indicators

The installed capacity of coal power under four pathways has decreased as the units are retired year by year. CCS transition can retain more coal power units for the system and provide more low-carbon and flexible installed capacity and coal power generation (see Fig. 23.2).

In 2060, the installed capacity of coal power under *Without CCS pathway* will be only 15 million kW (down 81% comparing with the starting year, accounting for 4% of the power structure). The larger the CCS transition scale, the higher the retained coal power capacity. Under high CCS pathway, the installed capacity of coal power will still be 50 million kW in 2060 (down 38% comparing with the starting year, accounting for 14% of the power structure), which is ten million kW and 20 million kW higher than the baseline CCS pathway and low CCS pathway, respectively. Among them, all the coal power retained under high CCS pathway has CCS function. The total installed capacity of wind and solar under high CCS pathway in 2060 will decrease by 82 million kW compared with 358 million kW under *Without CCS pathway*. Therefore, the development of coal power CCS can alleviate the construction pressure of new energy power generation under the same power supply guarantee responsibility.

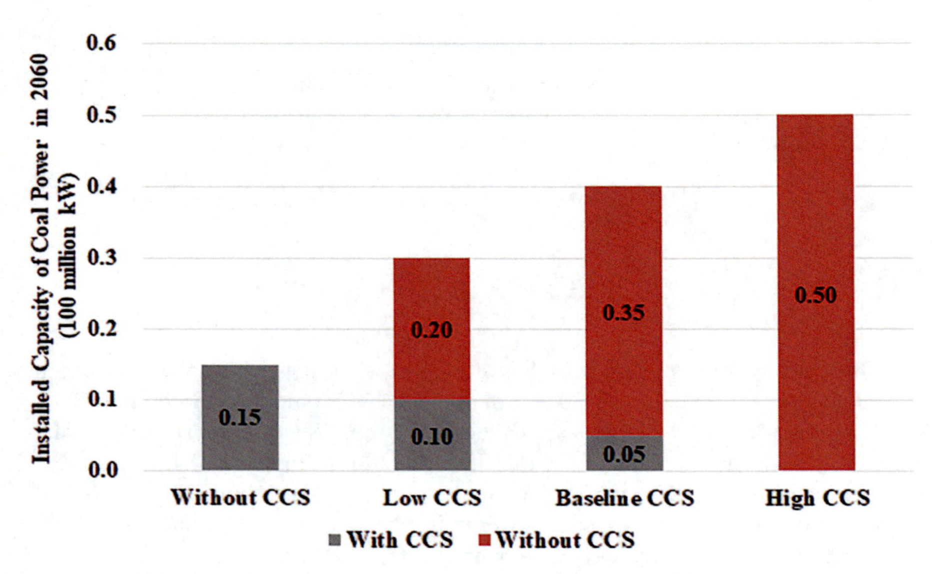

Fig. 23.2 Installed coal power capacity in 2060 under four pathways (with CCS and without CCS)

In terms of power generation, under the three *CCS development pathways*, coal power generation in 2060 will be 106–185 billion kWh, accounting for 13–23% of the total power generation, higher than the 6% of *Without CCS pathway* (see Fig. 23.3). The cumulative power generation of coal power in 2021–2060 will be 11.2–11.8 trillion kWh, 1.1–1.7 trillion kWh (11–17%) higher than that of the pathway without CCS.

The simulation results show that CCS can play an important role in making better use of abundant coal and coal power assets, reducing the adverse effects of intermittency and fluctuation of new energy, and optimizing the power structure.

23.3.2 Emission Indicators

Although the large-scale development of CCS has increased the power generation capacity of coal power and generated additional emissions, thanks to the high-carbon capture rate of CCS, the net emissions (carbon emissions generated by power generation minus captured carbon emissions) of each *CCS development pathway* are lower than those of the transition pathway without CCS.

In terms of annual emissions, the carbon emissions of four pathways will peak at 283 million tons in 2025. With the difference in *CCS development pathways* after 2030, the difference between the annual net emissions of high CCS pathway and other pathways will gradually widen, as seven million tons and 14 million tons lower than baseline CCS pathway and low CCS pathway in 2060 (see Fig. 23.4).

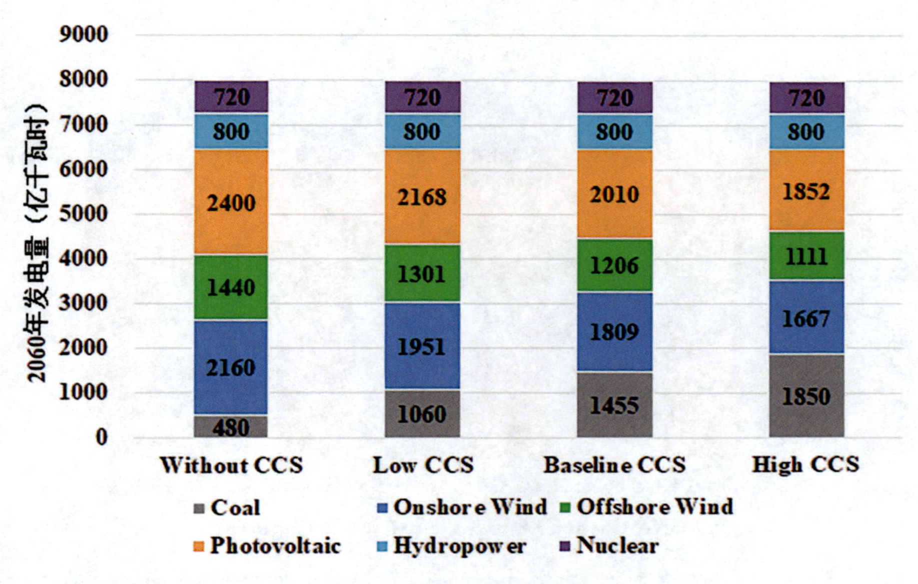

Fig. 23.3 Power generation structure in 2060 under four pathways

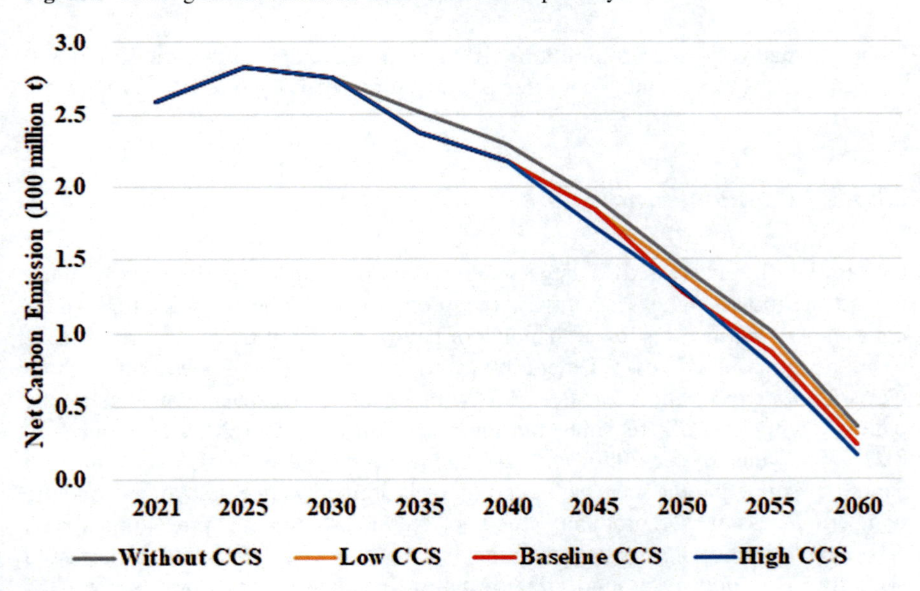

Fig. 23.4 Annual net carbon emissions under four pathways (2021–2060)

In terms of cumulative carbon emissions from 2021 to 2060, the cumulative total carbon emissions from 2021 to 2060 under high CCS pathway will be 9.8 billion tons, which is 300 million tons, 600 million tons, and 1 billion tons higher than the other three pathways, respectively. In terms of cumulative net carbon emissions,

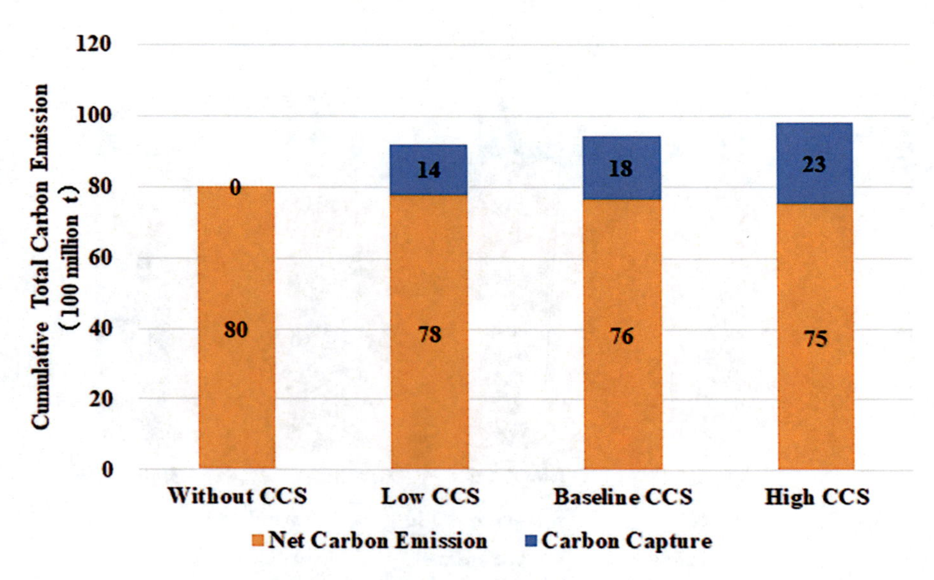

Fig. 23.5 Total cumulative carbon emissions from 2021 to 2060 under four pathways

high CCS pathway is 7.5 billion tons, 100 million tons, 300 million tons, and 500 million tons lower than the other three pathways, respectively (see Fig. 23.5).

23.3.3 Economic Indicators

Due to the high cost of CCS large-scale development in the early stage, if there are no additional policy incentives, the profit of coal power under three *CCS development pathways* is lower than that of pathway without CCS.

In terms of annual power generation profit, the profit of coal power under three *CCS development pathways* after 2031 will begin to be lower than that of the pathway without CCS, reaching the maximum difference in 2050 at about 6.2–8.2 billion yuan. After 2050, with the reduction of the cost of carbon dioxide capture per unit and the increase of carbon emission allowance price, the economic competitiveness of CCS gradually emerged, and the profit of coal power under each *CCS development pathway* begins to increase year by year, all of which surpassed *Without CCS pathway* around 2058. Among them, coal power profits grew the fastest under high CCS pathway, reaching 4.7 billion yuan in 2060, 2.7 billion yuan higher than the pathway without CCS (see Fig. 23.6).

In terms of the cumulative profit of coal power from 2021 to 2060, the higher the CCS development scale, the lower the cumulative profit. The cumulative profit of coal power from 2021 to 2060 under *Without CCS pathway* is 411.1 billion yuan,

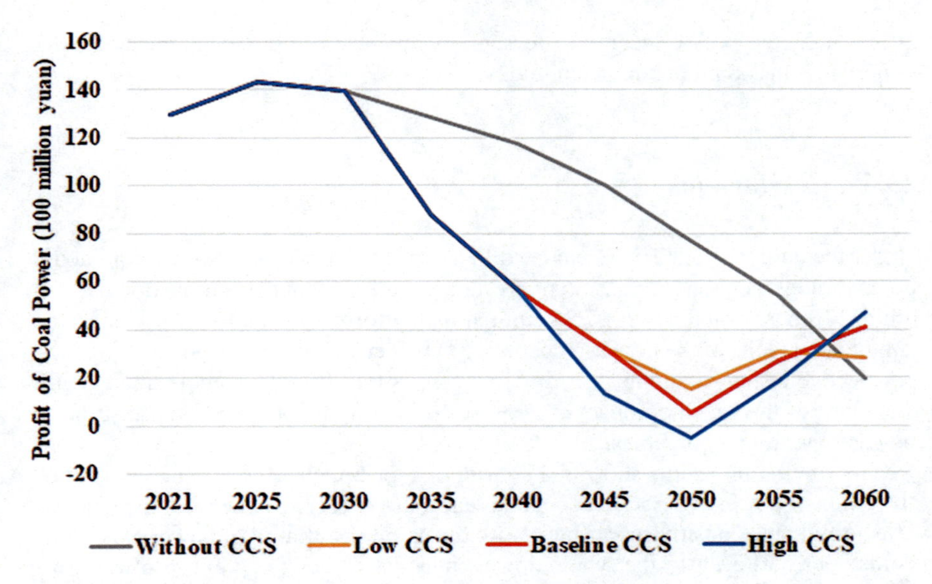

Fig. 23.6 Coal power profit under four pathways (2021–2060)

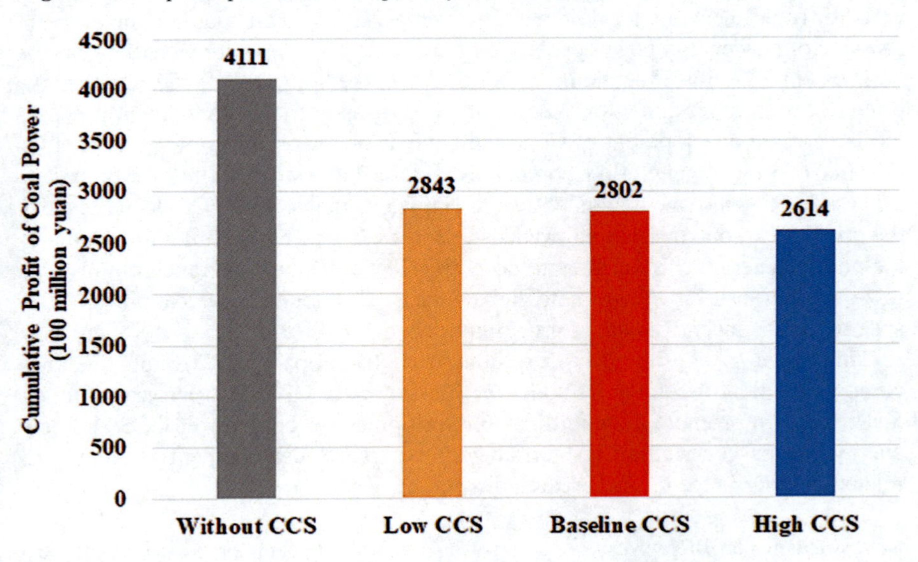

Fig. 23.7 Cumulative coal power profit from 2021 to 2060 under four pathways

which is 126.8–149.7 billion yuan higher than three *CCS development pathways* (see Fig. 23.7).

In order to increase the enthusiasm of power generation companies to develop CCS in the early and medium term of low-carbon transition, it is recommended to give targeted policy incentives to support the sustainable development of this important emission reduction technology.

23.4 Conclusions

Under the guidance of CPSSE framework, this paper establishes a simulation model of technology-economy-emission of power generation companies transition considering CCS based on the Sim-CPSS simulation platform; quantitatively evaluates the simulation results of the transition pathway of power generation companies from the perspective of power supply security, low-carbon transition, and economic benefit; and clarifies the specific impact of large-scale CCS on the low-carbon transition of power generation companies.

The simulation results show that considering CCS technology in the low-carbon transition of power generation companies can achieve significant emission reduction effects and make positive contributions to the carbon neutrality of power generation companies and even to the whole power industry. Under the premise of ensuring power supply, CCS technology can retain more low-carbon and flexible coal power units for future new energy-dominated power systems and alleviate the investment pressure of new energy power generation scale and system operation challenges to a large extent. In terms of economic benefits, due to the high cost of CCS and the low price of carbon emission allowances in the early stage of the development process, the large-scale development of CCS will have an adverse impact on the economic benefits of power generation companies. In the later stage of the development process, CCS will have a competitive advantage as the cost of CCS decreases and the price of carbon emission allowances increases. Therefore, in order to encourage the power generation companies to develop CCS in the early and medium term of low-carbon transition and promote the large-scale development of coal power CCS, specific policy incentives are of great importance.

This paper is a preliminary exploration of the impact of CCS on the low-carbon transition of companies under the framework of CPSSE. Appropriate policy incentives can promote the coordinated development of coal power CCS and new energy power generation. The optimization of CCS development strategy and related policies is the future research direction.

Acknowledgments This work was supported by the NARI Group Corporation Science and Technology Project of "Research on Coordinated development of carbon emission reduction and carbon sink enhancement."

References

1. International Energy Agency.: Energy and climate change. pp. 36–37 (2015)

2. Intergovernmental Panel on Climate Change.: IPCC Special Report 1.5°C. pp. 18–19 (2018)
3. International Energy.: Agency climate resilience. pp. 17–18 (2021)
4. Kang, C., Chen, Q., Xia, Q.: Carbon capture technology applied to power system and its changes. Automation Electr. Power Syst. **34**(1), 1–7 (2010)
5. Zhang, J., Zhang, L.: Preliminary discussion on the development of carbon capture, utilization and storage towards the goal of carbon neutrality. Thermal Power Generation. **50**(1), 1–6 (2021)
6. Hepburn, C., et al.: The technological and economic prospects for CO2 utilization and removal. Nature. **575**(7781), 87–97 (2019)
7. Academy of Environmental Planning, Ministry of Ecology and Environment., Wuhan Institute of Geomechanics., Chinese Academy of Sciences and China Agenda 21 Management Center.: China Carbon Dioxide Capture Utilization and Storage (CCUS) Annual Report (2021). pp 17–21 (2021)
8. Xue, Y., Yu, X.: Beyond smart grid-cyber-physical-social system in energy future. Proc. IEEE. **105**(12), 2290–2292 (2017)
9. Shu, Y., Xue, Y., Cai, B.: A review of energy transition analysis: elements and paradigms. Automation Electr. Power Syst. **42**(9), 1–15 (2018)
10. Xue, Y., Xie, D., Xue, F., Huang, J., Cai, B., Cai, L.: A cross-domain interactive simulation platform supporting the research of information-physical-social systems. Automation Electr. Power Syst. **46**(10), 138–148 (2022)
11. Wen, Y., Cai, B., Yang, X.: Quantitative analysis of China's Low-Carbon energy transition. Int. J. Electr. Power Energy Syst. **119**, 105854 (2020)
12. Cai, B., Xue, Y., Fan, Y.: Optimization on trans-regional electricity transmission scale of China's Western renewable energy base: the case study of Qinghai Province. E3S Web of Conferences. **143**(5), 2012 (2020)
13. Yang, X., Cai, B., Xue, Y.: Quantitative assessment of clean transition of GenCo considering other participants' generation investment. Energy Procedia. **156**, 23–27 (2019)
14. International Energy Agency.: Special Report on CCS in clean energy transitions. pp. 92 (2020)
15. Wei, N., Jiang, D., Liu, S.: Cost competitiveness analysis of CCS renovation of coal power plant of China Energy Group. Proc. CSEE. **40**(4), 1258–1265 (2020)

Chapter 24
Measuring the Financial Impact of Typhoon Due to Climate Change

Ji-Myong Kim

24.1 Introduction

24.1.1 Climate Change and Natural Disaster

Climate change is causing negative and serious damage in many industries and sectors. In the long term, such damage is representative of sea level rise and the spread of disease. Also, in the short term, extreme weather phenomena caused by climate change are representative. For example, there are heavy rains, floods, cold waves, heat waves, typhoons, and droughts. These extreme weather events are occurring all over the world and are increasingly coming as a threat to mankind. However, these extreme weather events have always threatened mankind since the beginning of human history. However, these extreme weather phenomena are changing patterns compared to the past, causing more negative effects on mankind than in the past.

This phenomenon caused by climate change is well demonstrated in the 5th Assessment Report (AR5) of the United Nations Intergovernmental Panel on Climate Change (IPCC). According to the report, if the current greenhouse gas emission continues, it is expected that it may exceed 2.0°. In addition, although various scenarios exist, in all scenarios, the global average temperature is projected to rise by 1.5 °C or more by 2100 [1]. In addition, studies related to global warming draw conclusions similar to those of the IPCC report and warn of the negative effects of a warming Earth [2, 3]. In common, it is expected that climate change of the earth will definitely occur due to global warming, and the occurrence of events such as extreme weather phenomena is expected to increase sharply in the warmed earth.

J.-M. Kim (✉)
Mokpo National University, Mokpo, South Korea

X. Wang (ed.), *Future Energy*, Green Energy and Technology ,
https://doi.org/10.1007/978-3-031-33906-6_24

In order to reduce the damage caused by these increasing natural disasters, many studies related to climate change and natural disasters have been conducted [4–10]. However, it is difficult to quantify the damage caused by natural disasters in previous studies. The reason is that natural disasters have a very close correlation with population and wealth, so it is very difficult to distinguish these factors from the damage caused by natural disasters.

Therefore, this study aims to quantitatively investigate the relationship between typhoons and climate change among the damage caused by natural disasters. To this end, by using the vulnerability function used in the natural disaster quantification model, factors such as population and wealth are excluded and the damage caused by natural disasters due to climate change is quantified.

24.2 Research Framework

The purpose of this study is to numerically identify changes in typhoons caused by climate change. To this end, the study was divided into two stages. In the first stage, changes in typhoons were analyzed by analyzing the intensity and frequency of typhoons in the past that had an impact in Korea by year. In the second step, the typhoon risk change was quantified using the Korean typhoon vulnerability function used in the catastrophic (CAT) model to quantify the risk due to climate change. Therefore, the scope of this study is limited to Korea.

24.3 Severity and Frequency of Typhoon by Decades

To analyze the severity and frequency of typhoons, the Korea Meteorological Administration (KMA) collected data on the number and maximum wind speed of typhoons that affected Korea from 1973 to 2019.

As shown in Table 24.1, the maximum wind speed of typhoons that affected South Korea during the period is organized by decade. As shown in Table 24.1, the maximum wind speed of typhoons affecting South Korea is increasing on average. Moreover, the coefficient of variation (CV) value is also increasing by decade, which shows that the deviation of large and small typhoons is increasing every year.

Table 24.1 Maximum wind speed of typhoons per decade

Dated	Ave.	STD.	CV
1970s	23.229	3.730	0.161
1980s	24.080	4.364	0.181
1990s	25.030	4.192	0.167
2000s	33.130	11.440	0.345
2010s	31.150	8.839	0.284

Table 24.2 Number of invading typhoons per decade

Dated	Ave.	STD.	CV
1970s	3.286	1.385	0.422
1980s	2.900	1.513	0.522
1990s	3.600	1.200	0.333
2000s	2.700	1.735	0.643
2010s	3.900	1.375	0.353

Table 24.3 Change in damage ratio by decade

Dated	Ave.	STD.	CV	Change rate
1970s	0.442%	0.283%	0.639	–
1980s	0.541%	0.339%	0.626	22.289%
1990s	0.566%	0.354%	0.625	28.025%
2000s	4.625%	3.457%	0.747	945.233%
2010s	2.821%	2.038%	0.723	537.417%

Table 24.2 shows the number of typhoons that affected Korea by decade. It does not show significant fluctuations by decade. Although the frequency of typhoons does not show a significant increase compared to the past, the intensity of typhoons increases every year, which shows that the risk of damage from typhoons increases.

24.4 Change in Damage Ratio by Decades

Table 24.3 shows the change in the damage ratio of buildings according to the change in typhoon wind speed by decade using the vulnerability function of the catastrophic (CAT) model in South Korea. As shown in Table 24.3, it is well shown that the damage ratio of buildings due to changes in typhoon wind speed for about 50 years from the 1970s to 2010 is rapidly increasing for every decade. The CV value is also increasing every decade, well showing that the variation according to the typhoon intensity is increasing. Moreover, the change rate shows the change in each decade based on the average damage rate in the 1970s, showing that it is rapidly increasing every decade. Therefore, it can be said that the damage rate of buildings is increasing rapidly due to climate change.

24.5 Discussion

In this study, changes in typhoons due to climate change were quantified. Changes in typhoons were analyzed by analyzing the frequency and intensity of typhoons that affected South Korea in the past. As a result of the analysis, the severity of typhoons is increasing rapidly compared to the past, and damage to buildings and people due to typhoons is expected to increase.

In addition, in this study, the risk of typhoon risk was investigated by numerically quantifying the risk of damage to buildings due to changes in the wind speed of typhoons. For scientific and objective identification, the vulnerability function of the CAT model that is not distorted by factors such as population or inequality was used. As a result of the analysis, the rate of damage to buildings due to typhoons increases significantly every decade, and a rapid increase in property damage is expected due to typhoons.

Quantifying the risk of damage from natural disasters due to climate change is an important part of both the private sector and the government. Therefore, it is necessary to look back on the increase in typhoon risk due to climate change through the research results and framework of this study and to derive various reduction and preventive measures for the increased risk.

24.6 Conclusions

Climate change due to global warming is a fact we face as seen in several studies. Therefore, research on natural disaster changes due to climate change should be continued. As seen in this study, typhoons are stronger than in the past, and as a result, typhoon damage to buildings is expected to increase rapidly. Therefore, active measures from the private sector and the government are required.

However, in this study, the typhoon records of about 50 years owned by KMA were collected and analyzed. Therefore, additional data collection is required for a more macroscopic analysis. In addition, since the research scope of this study is limited to Korea, it is difficult to directly apply it to other countries. In addition, since the scope of this study is limited to Korea, it is difficult to apply it directly to countries with different building codes or topography from Korea.

Acknowledgments This research was funded by Basic Science Research Program through the National Research Foundation of Korea (NRF) funded by the Ministry of Education (NRF-2022R1F1A106314111).

References

1. IPCC: Summary for policymakers. In: Climate Change 2014: Impacts, Adaptation, and Vulnerability. Part A: Global and Sectoral Aspects. Contribution of Working Group II to the Fifth Assessment Report of the Intergovernmental Panel on Climate Change, p. 1820. IPCC, Cambridge, UK (2014)
2. The World Bank: Turn Down the Heat: Confronting the New Climate Normal, p. 275. World Bank, Washington, DC (2013)
3. Dietz, S., Stern, N.: Endogenous growth, convexity of damages and climate risk: how Nordhaus′ framework supports deep cuts in carbon emissions. Econ. J. **125**, 574–620 (2015)
4. Bouwer, L.M.: Have disaster losses increased due to anthropogenic climate change. Bull. Am. Meteorol. Soc. **92**, 39–46 (2011)

5. Nordhaus, W.D.: The economics of hurricanes and implications of global warming. Clim. Chang. Econ. **1**, 1–20 (2010)
6. Miller, S., Muir-Wood, R., Boissonnade, A.: An exploration of trends in normalized weather-related catastrophe losses. In: Diaz, H.F., Murnane, R.J. (eds.) Climate Extremes and Society, pp. 225–347. Cambridge University Press, Cambridge, UK (2008)
7. Fengqing, J., Cheng, Z., Guijin, M., Ruji, H., Qingxia, M.: Magnification of flood disasters and its relation to regional precipitation and local human activities since the 1980s in Xinxiang, Northwestern China. Nat. Hazard. **36**, 307–330 (2005)
8. Schmidt, S., Kemfert, C., Hoppe, P.: Tropical cyclone losses in the USA and the impact of climate change: a trend analysis based on data from a new approach to adjusting storm losses. Environ. Impact Assess. Rev. **29**, 359–369 (2009)
9. Changnon, S.A.: Temporal and spatial distributions of wind storm damages in the United States. Climat. Chang. **94**, 473–482 (2009)
10. Chang, H., Franczyk, J., Kim, C.H.: What is responsible for increasing flood risks? The case of Gangwon Province, Korea. Nat. Hazard. **48**, 339–354 (2009)

Chapter 25
The Literature Intellectual Structure of System Dynamics on Waste

Elsa Rosyidah, Joni Hermana, and I. D. A. A. Warmadewanthi

25.1 Introduction

Solid waste management (SWM) in environmental engineering is one of the most important public policy concerns facing state and municipal governments as a result of the population growth and economic expansion of the twentieth century [1, 2]. Every year, about 2 x 10^9 t of municipal solid garbage was produced. The average daily output per individual is 0.74 kg. Waste production worldwide is anticipated to reach 2.59 x 10^9 t per year by 2030 and 3.40 x 10^9 t per year by 2050. At least 33% of the garbage produced is thought to not be handled in an environmentally safe way. Nearly 40% of the garbage produced is dumped in landfills, while 11% is treated via contemporary incineration and about 19% is recycled or composted. Governments are exploring sustainable waste disposal techniques as they become more aware of the dangers and expenses of dumpsites [3]. Most waste recycling initiatives have been discontinued, particularly during the COVID-19 pandemic, for a variety of reasons, including worries about the coronavirus spreading through the collected solid waste [4, 5]. Waste management can be understood with a system dynamics approach.

System dynamics (SD) is a method that examines complex systems' nonlinear dynamics and properties as well as the long-term consequences of strategy,

E. Rosyidah
Environmental Engineering Department, Universitas Nahdlatul Ulama Sidoarjo, Sidoarjo, Indonesia

Department of Environmental Engineering, Institut Teknologi Sepuluh Nopember, Surabaya, Indonesia

J. Hermana (✉) · I. D. A. A. Warmadewanthi
Department of Environmental Engineering, Institut Teknologi Sepuluh Nopember, Surabaya, Indonesia
e-mail: hermana@its.ac.id

X. Wang (ed.), *Future Energy*, Green Energy and Technology,
https://doi.org/10.1007/978-3-031-33906-6_25

approach, and environment. The interactions between the many variables of the SD model could be captured and interpreted via the feedback loops present. SD models are frequently used to simulate socioeconomic and environmental phenomena due to their dependable modeling capabilities and clear displays of the structure and effects [6, 7]. It has been demonstrated that System Dynamics (SD) is an effective method for examining the dynamic behavior of all processes among all relevant variables [8].

SD function is typically used to make the system easier to understand as a whole and to create a variety of relevant policy scenarios to control the system's dynamic evolution mechanism when examining the connections between a system's behavior over time and its underlying structure and decision-making guidelines [9, 10]. One of the many fields where SD applications have been used in the management of urban waste [11–13], managing construction waste [14], recycling of electronic components [15], managing packaging waste [16], and farming systems [17].

System dynamics has been widely used in the waste management field to forecast municipal solid waste [18], scheduling and directing of solid waste should be optimized [19], evaluation of landfill capacity, municipal solid waste production, and related cost control [20] reducing waste generated during construction and demolition [14], collection program for used portable batteries [21], assessment of source separation for municipal solid trash [8], given the expected generation of garbage [22]. However, prior research on the subject of system dynamics on waste was often limited to a single country and/or one specific field [23]. The system dynamics on waste haven't gained much attention despite producing a sizable visual map that is displayed on an international level year after year using information from various published research. There hasn't been any research that specifically addresses the substantial correlation between affiliation, scholars, and the influence of academic works.

One strategy for comprehending research is bibliometric analysis, in general. Bibliometrics of intellectual structure is a technique for analyzing and measuring academic publications that combine mathematical and statistical methods. Statistical methods known as bibliometrics of intellectual structure were used to analyze data on peer-reviewed publications, such as books, magazines, reports, reviews, and conference proceedings. Utilizing bibliometric methodologies, the relationship between the quantitative approaches and the subject of the investigation has been demonstrated frequently [24]. This study proposes a research question, what is the intellectual structure analysis of mapping and trends for system dynamics on waste research around the world? Using a bibliometric intellectual structure analysis, the purpose of this research was to map the current state of global research and future development trends of system dynamics on waste studies.

25.2 Research Method

The bibliometric intellectual structure analysis in this paper was conducted using a substantial literature database. This survey covers key phrases connected to system dynamics on waste studies to locate and classify related works in the global Scopus database. The Scopus database was used as the main information source for this article since academics trust it as a reputable source of scholarly publications.

The keywords "system dynamic" and "waste" were utilized in the title, abstract, and author keywords of this study to locate pertinent data in the Scopus database. Data mining was restricted to acquiring completely revealed data for an entire year. The ensuing search method was used in data mining, (TITLE-ABS-KEY ("system dynamic*") AND TITLE-ABS-KEY (waste)) AND EXCLUDE (PUBYEAR, 2022)) as of December 2022. 827 papers over the past 48 years—from 1973 to 2021—were uncovered in this process. The information from the Scopus results has been extracted at this point in the investigation and was in CSV file format [25].

A Scopus website feature called "Analyze Search Results" displays bibliometric data from specific literary genres. Using this technology, we assessed and visualized the academic, organizational, and national articles. The frequency of publications and citations, the distribution of sources and topics, and other statistics can all be monitored by authors using this tool [26–28].

The obtained documents were then subjected to co-occurrence analyses using VOSviewer version 1.6.16. This study creates a network of keyword maps for the conceptual research topics using VOSViewer. Additionally, it makes use of a fully systematic computing technique and keyword association analysis [29, 30]. Microsoft Excel was used to make straightforward tables and descriptive statistics. After that, the investigation's results were compiled and confirmed.

25.3 Result and Discussion

25.3.1 Organizations and Countries with the Highest Productivity in System Dynamics on Waste Research

There were 827 articles affiliated with 1535 research institutions. The most productive organization in researching system dynamics on waste publications was the Chinese Academy of Sciences, China (n = 16). Then followed by Riga Technical University, Latvia (n = 16); Hong Kong Polytechnic University, Hong Kong (n = 14); Shenzhen University, China (n = 13), Southwest Jiaotong University, China (n = 13); Institut Teknologi Sepuluh Nopember, Indonesia (n = 11); Tianjin University, China (n = 10); and Università Degli Studi di Napoli Federico II, Italy (n = 9). System dynamics on waste research were dominated by affiliates from China (n = 4), Latvia (n = 1), Hong Kong (n = 1), and Indonesia (n = 1). These countries were active in research and publications in system dynamics on waste [31, 32].

These 79 countries were identified to have researched system dynamics on waste. The most productive country in researching system dynamics on waste publications was China (n = 225). Then followed by the United States (n = 115); Indonesia (n = 39); Italy (n = 39), United Kingdom (n = 39); Germany (n = 38); Australia

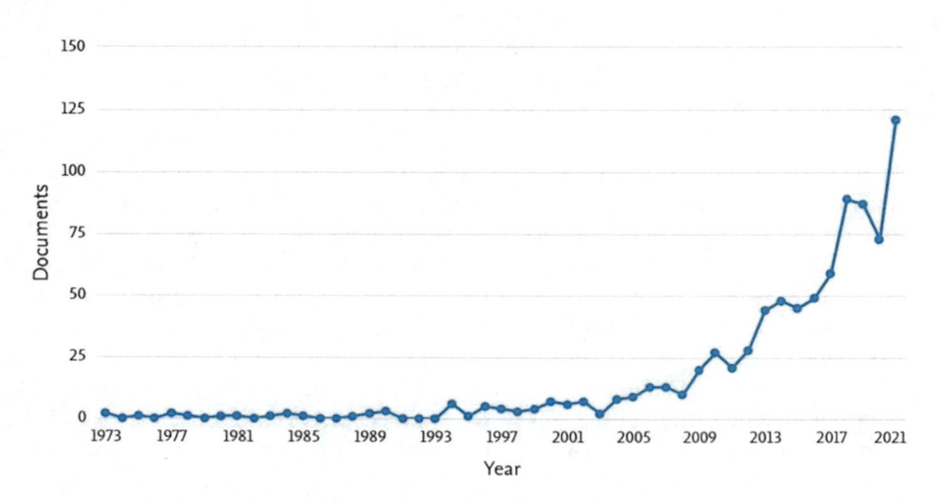

Fig. 25.1 The System Dynamics on Waste Sector's Annual Publications

(n = 31); Canada (n = 31); India (n = 31); Brazil (n = 27); Taiwan (n = 24); Japan (n = 19); Colombia (n = 18); Hong Kong (n = 18); and Netherland (n = 18).

Most system dynamics on waste publications were dominated by developed countries. The most productive countries in the publication of system dynamics on waste were supported by the productive institution.

25.3.2 Annual Publications for the System Dynamics on Waste Sector

There were more international articles on system dynamics in waste studies published each year. With 121 publications, Fig. 25.1 shows that 2021 have be the year of publishing's highest peak. According to the Scopus record, scientists have been researching waste projection since 1973. In general, based on the number of annual publications, system dynamics on waste research shows an upward trend in growth.

25.3.3 Research Map by Theme

Based on previously disclosed keyword correlations, the research theme map was a search for system dynamics on waste research. Figure 25.2 below shows the concept of the research theme in system dynamics on waste.

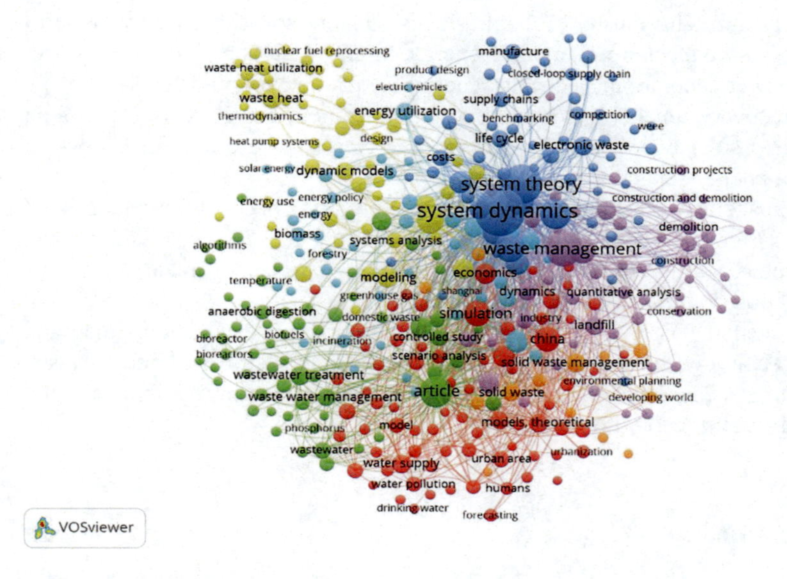

Fig. 25.2 Map of Research Themes

The system dynamics on the waste keyword scheme were examined and presented using the VOSViewer application for the research topic map's system dynamics on waste. The bare minimum of keyword-related articles required six repeats. Out of 373 keywords, 6623 were determined to satisfy the requirements. As a concept for GEMWEWS research themes, seven research subject groupings for international academic publications on waste projection have been streamlined and condensed based on research keywords.

1. Government clusters. This group of themes focuses on the government. The keyword of developing countries, developing world, hospital, and public health. Government components dominated this cluster. Numerous of these keywords relate in some way to the government.
2. Environment clusters. This cluster of themes focuses on the environment. The keyword of agriculture, climate change, ecosystem, environment, environmental economics, environmental sustainability, environmental protection, environmental pollution, environmental policy, human, pollution, river, sanitation, and water. This cluster was associated with the majority of these phrases.
3. Modelling clusters. We can find modeling topics in this research cluster. The keyword of computer simulation, design, dynamic analysis, dynamic modelling, dynamic model, mathematical models, model validation, modelling, models, numerical model, ranking cycle, system dynamic models, and technology.
4. Waste clusters. The keyword of waste disposal, waste treatment, wastewater, wastewater management, wastewater treatment plant, wastewater, and wastewater treatment dominated the water cluster.

5. Energy clusters. This cluster of themes focuses on energy. These keywords form a cluster that connects to alternative energy, biomass, carbon, carbon emission, electric power generation, electricity, energy, energy consumption, energy policy, energy recovery, energy resource, energy utilization, fossil fuels, renewable energy, and solar energy that dominated this cluster. These keywords cover a variety of energy-related subjects.
6. Waste management clusters. The keyword of conservation, construction, decision support system, decision support tools, environmental management, project management, waste reduction, waste generation, waste management, and waste minimization dominated the waste management cluster.
7. System clusters. The keyword of assessment method, causal loop diagram, circular economy, closed-loop, conceptual model, decision making, e-waste, environmental planning, life cycle, system dynamics, system dynamic simulation, and system theory dominated the system cluster.

25.4 Conclusion

According to data, there have been more worldwide publications on system dynamics on waste, maps, and visual trends each year. The most active research institution and nations were the Chinese Academy of Sciences and China, with 16 and 225 publications published in the system dynamics publications on waste. 2021 had the most academic publications, with 121 papers on the topic of system dynamics on the waste study published globally. System dynamics on waste research show an upward trend in growth based on the number of annual academic publications.

The GEMWEWS research themes—government, environment, modeling, waste, energy, waste management, and system—were used in this study to classify the body of knowledge created over forty-eight years of scholarly publication in terms of knowledge contributions. Practical field studies are essential to pinpoint broad backgrounds and disciplines, as well as research gaps, as a result of defining major concerns in system dynamics in the waste industry and a better understanding of their demand. The lack of advanced knowledge, assessment, and analysis in the disciplines will be the subject of new research made possible by all of this. Research on sustainability, literature, and system dynamics on waste's potential to enhance decision-making in environmental protection and management. Authors recommend to use Vensim PLE to run the system dynamic model to simulate the activities of waste management system.

Acknowledgments The authors would like to thank the Ministry of Education, Culture, Research, and Technology (Kemdikbudristek) of the Republic of Indonesia for the Doctoral Dissertation Research Grant (PDD).

References

1. Pinha, A.C.H., Sagawa, J.K.: A system dynamics modelling approach for municipal solid waste management and financial analysis. J. Clean. Prod. **269**, 122350 (2020). https://doi.org/10.1016/j.jclepro.2020.122350
2. Hermana, J., Nurhayati, E.: Removal of Cr3+ and Hg2+ using compost derived from municipal solid waste. Sustain. Environ. Res. **20**(4), 257–261 (2010)
3. Kaza, S., Yao, L., Bhada-Tata, P., Van Woerden, F.: What a Waste 2.0 : a Global Snapshot of Solid Waste Management to 2050. Urban Development, Washington, DC (2018)
4. I. Warmadewanthi *et al.*, "Socio-Economic Impacts of The COVID-19 Pandemic on Waste Bank Closed-Loop System in Surabaya, Indonesia," *Waste Manag.* Res. J. a Sustain. Circ. Econ., vol. 39, no. 8, pp. 1039–1047, Aug. 2021, https://doi.org/10.1177/0734242X211017986
5. Wulandari, D., et al.: Solid waste Management in a Coastal Area (study case: Sukolilo sub-district, Surabaya). IOP Conf. Ser. Earth Environ. Sci. **799**(1), 012030 (2021). https://doi.org/10.1088/1755-1315/799/1/012030
6. Guan, D., Gao, W., Su, W., Li, H., Hokao, K.: Modeling and dynamic assessment of urban economy–resource–environment system with a coupled system dynamics – geographic information system model. Ecol. Indic. **11**(5), 1333–1344 (2011). https://doi.org/10.1016/j.ecolind.2011.02.007
7. Liu, H., Benoit, G., Liu, T., Liu, Y., Guo, H.: An integrated system dynamics model developed for managing Lake water quality at the watershed scale. J. Environ. Manag. **155**, 11–23 (2015). https://doi.org/10.1016/j.jenvman.2015.02.046
8. Sukholthaman, P., Sharp, A.: A system dynamics model to evaluate effects of source separation of municipal solid waste management: a case of Bangkok, Thailand. Waste Manag. **52**, 50–61 (2016). https://doi.org/10.1016/j.wasman.2016.03.026
9. Yuan, X.H., Ji, X., Chen, H., Chen, B., Chen, G.Q.: Urban dynamics and multiple-objective programming: a case study of Beijing. Commun. Nonlinear Sci. Numer. Simul. **13**(9), 1998–2017 (2008). https://doi.org/10.1016/j.cnsns.2007.03.014
10. Peck, S.: System enquiry: a system dynamics approach. J. Oper. Res. Soc. **42**(10), 906–907 (1991). https://doi.org/10.1057/jors.1991.175
11. T. Wei, I. Lou, Z. Yang, and Y. Li, "A system dynamics urban water management model for Macau, China," J. Environ. Sci., vol. 50, pp. 117–126, Dec. 2016, https://doi.org/10.1016/j.jes.2016.06.034
12. Giannis, A., Chen, M., Yin, K., Tong, H., Veksha, A.: Application of system dynamics Modeling for evaluation of different recycling scenarios in Singapore. J. Mater. Cycles Waste Manag. **19**(3), 1177–1185 (2017). https://doi.org/10.1007/s10163-016-0503-2
13. Di Nola, M.F., Escapa, M., Ansah, J.P.: Modelling solid waste management solutions: the case of Campania, Italy. Waste Manag. **78**, 717–729 (2018). https://doi.org/10.1016/j.wasman.2018.06.006
14. Ding, Z., Yi, G., Tam, V.W.Y., Huang, T.: A system dynamics-based environmental performance simulation of construction waste reduction management in China. Waste Manag. **51**, 130–141 (2016). https://doi.org/10.1016/j.wasman.2016.03.001
15. Fan, C., Fan, S.-K.S., Wang, C.-S., Tsai, W.-P.: Modeling computer recycling in Taiwan using system dynamics. Resour. Conserv. Recycl. **128**, 167–175 (2018). https://doi.org/10.1016/j.resconrec.2016.09.006
16. Dace, E., Bazbauers, G., Berzina, A., Davidsen, P.I.: System dynamics model for Analyzing effects of eco-design policy on packaging waste management system. Resour. Conserv. Recycl. **87**, 175–190 (2014). https://doi.org/10.1016/j.resconrec.2014.04.004
17. Saysel, A.K., Barlas, Y., Yenigün, O.: Environmental sustainability in an agricultural development project: a system dynamics approach. J. Environ. Manag. **64**(3), 247–260 (2002). https://doi.org/10.1006/jema.2001.0488

18. Dyson, B., Chang, N.-B.: Forecasting municipal solid waste generation in a fast-growing urban region with system dynamics Modeling. Waste Manag. **25**(7), 669–679 (2005). https://doi.org/10.1016/j.wasman.2004.10.005
19. Johansson, O.M.: The effect of dynamic scheduling and routing in a solid waste management system. Waste Manag. **26**(8), 875–885 (2006). https://doi.org/10.1016/j.wasman.2005.09.004
20. Kollikkathara, N., Feng, H., Yu, D.: A system dynamic Modeling approach for evaluating municipal solid waste generation, landfill capacity and related cost management issues. Waste Manag. **30**(11), 2194–2203 (2010). https://doi.org/10.1016/j.wasman.2010.05.012
21. Blumberga, A., Timma, L., Romagnoli, F., Blumberga, D.: Dynamic modelling of a collection scheme of waste portable batteries for ecological and economic sustainability. J. Clean. Prod. **88**, 224–233 (2015). https://doi.org/10.1016/j.jclepro.2014.06.063
22. Johnson, N.E., et al.: Patterns of waste generation: a gradient boosting model for short-term waste prediction in new York City. Waste Manag. **62**, 3–11 (2017). https://doi.org/10.1016/j.wasman.2017.01.037
23. Lee, C.K.M., Ng, K.K.H., Kwong, C.K., Tay, S.T.: A system dynamics model for evaluating food waste Management in Hong Kong, China. J. Mater. Cycles Waste Manag. **21**(3), 433–456 (2019). https://doi.org/10.1007/s10163-018-0804-8
24. IGI Global: What Is Bibliometric? IGI Global (2021) [Online]. Available: https://www.igi-global.com/dictionary/education-literature-development-responsibility/2406
25. Rosyidah, E.: The Literature Intellectual Structure of System Dynamics on Waste Dataset (1973–2021). Mendeley Data (2022)
26. A. Purnomo, N. Asitah, E. Rosyidah, H. Ismanto, and R. A. M. A. Lestari, "Green technology: lesson from research mapping through bibliometric analysis," IOP Conf. Ser. Earth Environ. Sci., vol. 1063, no. 1, p. 012022, Jul. 2022, https://doi.org/10.1088/1755-1315/1063/1/012022
27. Purnomo, A., Sari, A.K., Aziz, A., Prasetyo, Y.E., Rosyidah, E.: A study of green management literature through bibliometric positioning during four decades. In: 11th Annual International Conference on Industrial Engineering and Operations Management Singapore. IEOM, Singapore (2021)
28. Purnomo, A., Septianto, A., Rosyidah, E., Khan, H.A.U., Purnama, P.A.: Green manufacturing literature during three decades: a Scientometric approach. IOP Conf. Ser. Earth Environ. Sci. **729**(1), 012046 (2021). https://doi.org/10.1088/1755-1315/729/1/012046
29. Ranjbar-Sahraei, B., Negenborn, R.R.: Research Positioning & Trend Identification. TU Delft, Walanda (2017)
30. van Eck, N.J., Waltman, L.: Software survey: Vosviewer, a computer program for bibliometric mapping. Scientometrics. **84**(2), 523–538 (2010). https://doi.org/10.1007/s11192-009-0146-3
31. Xiao, S., Dong, H., Geng, Y., Tian, X., Liu, C., Li, H.: Policy impacts on municipal solid waste Management in Shanghai: a system dynamics model analysis. J. Clean. Prod. **262**, 121366 (2020). https://doi.org/10.1016/j.jclepro.2020.121366
32. Galli, F., Cavicchi, A., Brunori, G.: Food waste reduction and food poverty alleviation: a system dynamics conceptual model. Agric. Human Values. **36**(2), 289–300 (2019). https://doi.org/10.1007/s10460-019-09919-0

Index

X. Wang (ed.), *Future Energy*, Green Energy and Technology,
https://doi.org/10.1007/978-3-031-33906-6

U

V

W

Printed in the United States
by Baker & Taylor Publisher Services